Published by Cornwall Stargazers

Printed in Great Britain

ISBN 9781973529125

Contents

Acknowledgements

The author wishes to thank the many individuals who have supported this research. This book would have not been possible without the enthusiasm and patience of my family, particularly my husband Eddie who never questioned my late night and early morning visits to the stone circle sites, and my Father for proof reading this book. Brian Sheen of the Roseland Observatory has been a mentor and friend throughout the project. Without his assistance and expertise this book would not have been possible. Also, much appreciation goes to Cheryl Straffon who kindly provided guidance to the west Penwith sites and insightful discussions which have shaped the content of this book.

Preface

Stone circle sites are enigmatic in their nature. Links between the circle sites throughout the UK have been made to celestial events, particularly events linked to the Sun, Moon and stars. Current work in the field takes a holistic view of sites and their surroundings considering how people would have related to sky, landscape and how they may have translated these ideas into monuments which remain in our environment today.

This book explores astronomical connections at four such sites all located within the Penwith region of West Cornwall. The sites are four remaining stone circles called; Boscawen-ûn, The Merry Maiden, Boskednan and Tregeseal. The book is the culmination of several years' work within the region. It initially looks at the types of alignments which can be made with the stone circles, and considers if they could have been observed and placed by people of the Bronze age with celestial events taken into consideration. Each of the circles are examined in turn with landscape features and the location of each circle being described. Historic archaeoastronomy suggestions have been assessed to see if they hold true today. If any myths and legends are known about the sites these have equally been explored.

Each site has then had a new archaeoastronomy survey conducted at them, the types of alignments examined are primarily solar, lunar, but stellar connections are also suggested. Conclusions have been drawn about links to the celestial motions of the sky and if the circles have been positioned in the landscape with astronomy in mind. I hope you enjoy reading this book as much as I have had researching and writing it.

Introduction

West Penwith is a landscape where there are numerous historic stone monuments, positioned by people deep in the past. Neolithic and bronze age monuments are densely spaced here. These stones are evocative, redolent, steeped in myths and legend. Tales involve people being turned to stone for dancing and playing instruments and these are reflected in the names given to them "The Pipers", "Nine Maidens" and the "Merry Maidens". It's also been suggested that medical cures could be found at a rocking stone of Nancladra, or a holed stone of Mên-an-Tol which would cure those which crawled or were passed through it. One such stone called the "Creeping Stone" would cure you if you crept towards it against the Sun. The astronomical links do not stop there, the Quartz stone at Boscawen-ûn is considered symbolic of the Moon. Legends involving celestial features aren't common but do reflect a continuing fascination with man and the movement of astronomical objects in the sky. Nevertheless, the stones which are left in the landscape have been difficult for us, as modern people, to decipher and understand. It has been suggested many times over, the people who placed these stones in our landscape did so with an eye turned skywards. They created a landscape which in many ways reflected the skyscape, with abundant sites being astronomically aligned to events in the sky. Documented through history are numerous suggestions of how the skyscape has been reflected in the design and positioning of the stones. This study of archaeoastronomy has led to a number of theories surrounding megalithic sites and their existence. Their differing interpretation covers a wide range of ideas. These ideas include suggestions of; eclipse predictors, time-keepers and azimuth confirmation to name a few. The range of megalithic sites covered is equally as diverse. They consider stone circles, barrows, and menhirs. One question which has often cropped up is why? what is the purpose of alignments? The reasons again, have covered many aspects of megalithic life, including religion, agricultural, political strength obtained through predictive powers or even pure curiosity. Over time some sites, such as Stonehenge, have had numerous archaeoastronomy theories placed upon them, others such as Newgrange have had more focused theoretical predictions.

One particular monument; the stone circle deserves consideration, as they have evoked countless astronomical suggestions. Some of these suggestions hold value, but others are less easy to reconcile. This book will address each of the four remaining stone circle examples in West Penwith and consider if they have been astronomically aligned. It will examine how they are placed in the landscape and see if there is a wider reflection of astronomy through the landscape. For each of the remaining circles the historic archaeoastronomy suggestions will be considered with a critical eye. Following this new alignment measurements will be made using modern techniques and it will be considered if there were any astronomical alignments in the design of the remaining West Penwith circles.

The wider environment

West Penwith is located in the far southwest of the United Kingdom within the Duchy of Cornwall. This region forms the profile of the toe on the foot. The name derives from two Cornish words, penn meaning 'headland' and wydh meaning 'at the end'. It is attached to the mainland in the west by a six-mile strip of land, which runs approximately between the towns of St Ives in the north and Penzance in the south. The north, west and south sides are banded by the sea, the coastline is rugged, with few beaches and small natural harbours, often reached through deep valley cuts. The land is a granite outcrop. It has two distinct halves, a southern plateau rising about 100m above sea level and a northern moorland area, which rises to above 200m. The area has a number of dramatic and well-known Cairn's including Chapel Carn Brea, Carn Galver, Carn Kenidjack and Bartinney.

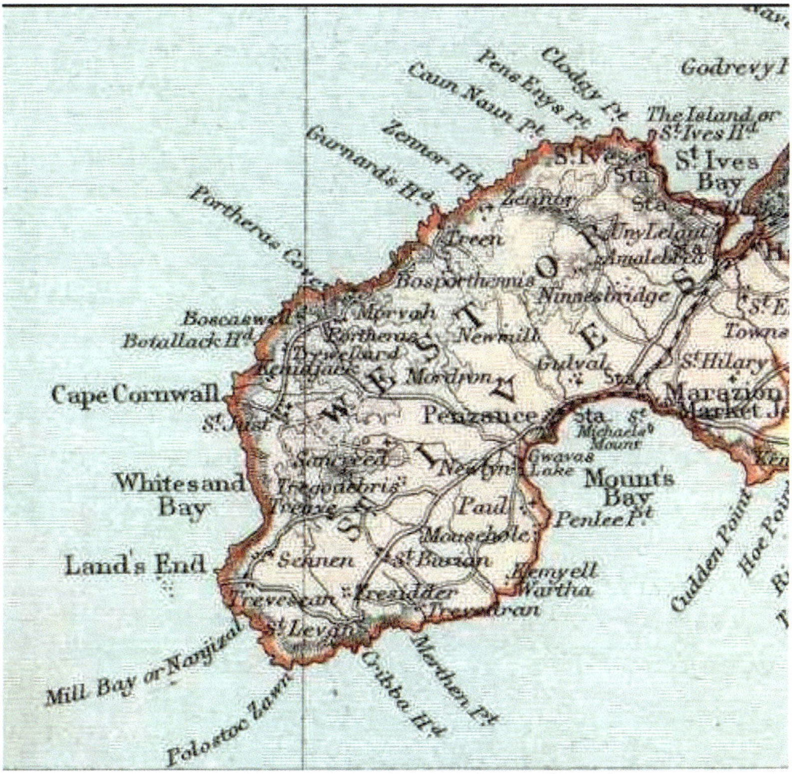

Map of West Penwith – Mackenzie (1895).

It is an isolated and rural region of the country, which still contains a large concentration of prehistoric remains within the landscape. Evidence of human activity dates back to the time of the reoccupation of Britain after the last Ice Age. Although much of the area would have been originally wooded, there is no evidence to suggest that this woodland survived the influx of human agricultural practices and settlement in the early Neolithic period. By the late Neolithic much of West Penwith was occupied, with Neolithic field systems being visible in a number of locations including around the remaining stone circles. The far-reaching views that are visible today would be similar during this period. The Neolithic period brought evidence of the first mines. Importantly the area is rich in tin, which is essential in the formation of bronze. By the early Bronze Age, there were numerous mines and trading routes in the area. Greenstone axes which originate from West Penwith are distributed throughout southern England. Early trading tracks, such as the local Tinner's way would be well used routes for movement of both people and goods. This trade would have extended long distances over the sea as well as land, with influence and artefacts found in the region from the rest of England as well as Brittany and Ireland. The discovery of the Nebra sky disc (dated 1600BCE) in Germany using Cornish tin and gold show how far raw and valuable metals could be traded during this period. Large stone monuments such as stone circles, menhirs and barrows were built on the peninsula until about 1200BCE, when for some reason they ceased being built.

Stone circles in West Penwith

Stone circles are evident throughout the peninsula from the period. Precise dating of stone circles is difficult, the oldest stone circle enclosures within the UK date from 3300BCE, with early parts of Stonehenge dating from 3200BCE. John Barnatt in Prehistoric Cornwall (1982) suggests that the majority of stone circles in Cornwall date between 2500 - 1600 BCE, although more recent archaeological discussion places them between 1900 and 1200BCE. Many of the stone circle sites would have had an evolution of development. The circle sites would have very possibly been of spiritual value before the stone monument was built. Therefore, while considering alignments a date of 2500BCE will be used through this project, as this links what is the earliest date as suggested by Barnatt to the evolving development of the sites. This project considers alignments to natural and manmade features. The natural alignments are considered as a reason for the positioning of the circle in their location.

Four near-complete stone circles remain. There is evidence for a total of fifteen having been built. The four remaining circles are Tregeseal East near St Just, Boscawen-ûn just off the A30 near Crows-an-wra, The Merry Maidens found just south of Lamorna and Boskendan high on the moor south of Zennor. Near Boskendan there is the better known but dilapidated Mên-an-Tol. It is famous for its large holed stone which is considered to have been part of a circle. The evidence for other circles come from a number of sources including; first hand historic descriptions, historic drawings, crop circle markings and old ordnance survey maps.

A number of the circles were built as multiple circle sites, such as at Tregeseal where there existed up to three circles. Also, the Merry Maidens site had two circles side by side and a possible third a short distance away. At Boscawen-ûn there is evidence for only one circle. The circles are located throughout the peninsula, in both the highland northern half of peninsula as well as the lowland south. You can find then by the coastline as well as within the interior. The overall size of the circles varies, although there are no large circles, such as found on Bodmin moor. The layouts are both irregular and regular circles, with Boscawen-ûn classified by the famous archaeoastronomers Alexander Thom as a flat circle shape and the Merry Maidens being almost circular. There is evidence of the grading of the stones in height order at some circles such as the Merry Maidens and Boskednan, although this does not seem to be the case at Boscawen-ûn. Only Boscawen- ûn has a central stone. All the stone circle sites have outliers such as standing stone (menhirs) and barrows in close proximity. Some outliers form straight rows leading from the circles across the landscape. A number of the circles have had astronomical alignments suggested, particularly solar alignments such as the solstice and equinox Sunrise and Sunset. These alignments tend to be with features on the horizon.
Each of the four-remaining circle's will be examined in turn with a detailed look at its design and position in the landscape. Any historic astronomical suggestions made about alignments

at the circle will outlined and discussed. But first the type of alignments that can be made at each circle will be outlined in the next chapter.

Alignments

When setting off on a project such as this the types of celestial events we wish to align needs to be considered. This section will look at these events. It will examine if and how often these event alignments can occur and if they are a possibility for alignment at the stone circles sites.
The choice of astronomical objects can be wide ranging, and events could include solar, lunar, planetary, stellar and events such eclipse, comets, supernovae or aurora. When drawing up a list considering alignments with the monuments we have to understand the types of celestial events that may have been important to the ancient society and they may have marked. It is very difficult to turn back the clock 4000 years to consider what they may have considered meaningful. But a way to start this process is to look at other sites which have alignment examples, or through links to the past through myths and legends. Once we have decided what astronomical features we would like to measure, we also need to understand what is feasible for us to measure. We cannot suggest alignments to objects where the positions would be difficult for us to reconstruct. The events will be examined in turn with solar, lunar, planetary and stellar events discussed below.

The Sun

The Sun is the most important celestial object known to humans. It brings daily warmth and light to an otherwise cold and dark world. There is no wonder that it has been observed and revered since the earliest of times. It is no surprise therefore that the motion of the Sun was well understood in the bronze age, but it is worth describing what happens and can be viewed in the northern hemisphere. The Sun would rise daily in the east and set daily in the west. At a UK latitude of 50 degrees it rises directly in the east at approximately 6am on the morning of spring equinox before setting at approximately 6pm twelve hours later. On the equinox the Sun at noon is directly over the Earth's equator. Each day the Sun moves towards the summer solstice its rising and setting position displaces daily along the horizon towards the north. It rises early and sets later until we reach the midsummer solstice point. At this point the Sun at noon is directly over the tropic of cancer. In the northern hemisphere this means that the Sun rises and sets at its most northerly azimuth on the summer solstice. It also rises to its highest position in the sky, rising and setting quickly with its greatest angle to the horizon. After the summer solstice the amount of daylight hours one more start to reduce, as the Sun returns to the equinox position but this time the autumnal one. Its rising and setting position continue to move south along the horizon. When it reaches its most southerly position on the horizon it has reached the winter solstice. The daylight hours are short and the angle that the Sun rises and sets from the horizon is at its lowest. At this point the Sun at noon is over the tropic of Capricorn. A view of the rising and setting points can be seen in the diagram below.

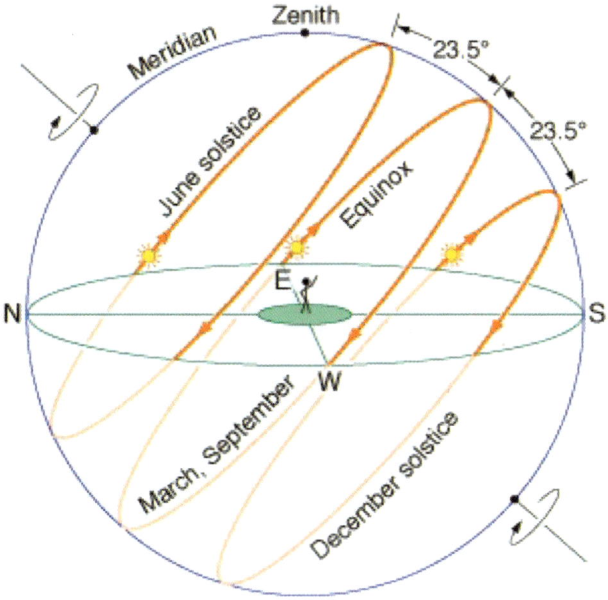

Rising and setting points of the Sun through the solar year
Credit Daniel Schroeder (2016)

The position that the Sun rises and sets can be ascribed a value, This value is known as a declination. Declinations are a measurement of position attributed all celestial objects, it works in similar way to latitude. Just as latitude can help describe how far north or south you are on Earth, declination describes the position north or south on the celestial equator on the celestial sphere. The current declination of the Sun has a range of +/- 23.5° either side of the equinox (mid-point). This declination range is accurate for the current epoch and echo's how Earth's path is currently tilted by 23.5°with respect to the Sun. This tilt is due to the angle between the Earth's orbital plane and its rotational axis and is called the obliquity of the ecliptic or ε. The value of ε isn't static, there are small changes in it over time due to the changing tilt of the Earth, this in turn changes the declination range of the Sun. It means that the rising and setting position of the Sun on the horizon has changed since Neolithic times. To work out the amount of difference we have to recognise the change in the tilt of the Earth and therefore by how much the declination range of the Sun has moved. The current and past values of ε can be seen in the table. They have been reconstructed for the period that we are interested in.

3000BC	24° 01' 15.050"
2500BC	23° 58' 26.924"
2000BC	23° 55' 26.026"
1600BC	23° 52' 52.959"
2016 current	23° 26' 13.6"

Now we recognise that the declination of the Sun has changed, we must understand what that means to the rising and setting position of the Sun. At the equinox the Sun would rise and set almost in the same position as it always has throughout history (East and West). It would be at the solstices where this change in the declination would make the most difference to the rising and setting position of the Sun. Following shows how much difference this would make for the rising and setting Sun in West Cornwall between the years of 2500BCE and 2016AD. The central position of the Sun rising with a declination of 23°25'57" on the solstice 2016. The central position of the rising Sun in 2500BCE has a declination of 23.9°. If we were to look at the solar centre in 2016 at an altitude 0°it has an azimuth of 308°18'.03". The Sun at 2500BCE at an altitude 0° shows an azimuth of 309°20'. This gives a change of 1°02', therefore, at its extreme positions the Sun would rise in 1.02°towards the north at the summer solstice 2500BCE and set 1.02° further towards the south at the winter solstice in 2500BCE more than it does currently. The Sun's motion along the horizon can aid people to pinpoint the extreme points of its cycle. The Sun slows its daily displacement along the horizon as it approaches the solstice before changing its path along the horizon again.[1] On the approach to solstice, it takes 11 days for the Sun to have moved a whole solar diameter. At solstice +/- 1 day it only moves 2', at +/- 7 days the daily change of declination reaches 3', whereas at the equinox the daily change is 24' or nearly a whole solar width in a day. This means that at the Sun approaches the solstice it appears to rise and set in the same position for a number of days. This means that the extreme values or solstice are considered a key part of the cycle to measure. There are many local sites which have known solstice alignments these include Craddock Moor and the Triplet Stone circles on Bodmin Moor Cornwall.

It is not just the solstice that was important, other solar dates are worthwhile making observations on too. This includes the equinox (midpoint between the two solstices). These are also the cardinal points of east and west at any given site and are located at 90°and 270°.

The position of the equinox has not changed due to ε and would always have been observed at 90° and 270° throughout history. At the equinox the daily displacement of the Sun reaches its maximum with a daily movement of 24'. Some scientists consider that equinox alignments would have been unlikely. But others dispute this and believe that there are enough examples of equinox alignments to make them viable. In Cornwall examples of equinox alignments can be seen at the Hurlers, Leskernick and Goodaver stone circles on Bodmin Moor.

Cross quarter days fall between the solstice and equinox, the year is split into 8 equal parts. They are listed in the table below with the Celtic names they are often associated with. Celtic festivals have been documented back to the Iron age, but they could have their roots much further back. Local cross quarter alignments are suggested for Stannon and Goodaver stone circles on Bodmin Moor. The declination of the Sun can be calculated with,

$$Sin\delta = sin\,\varepsilon\, cos\,(0.9856n)$$

Where n = the number of days since the June solstice, (there would be some difference due to Earth's motion around the Sun being an ellipse, but this would be minor, Also the figures are for the horizon and for instance the equinox point could fall above and below this point, leading to a small amount of difference in the declination than the figure given below).

[1] *It isn't though considered reasonable or possible to pinpoint the actual moment of solstice through observation due to refraction and atmospheric conditions at the horizon.*

Key dates	Declination	Sunrise Az	Sunset Az	Name	Key dates	Declination
2016AD	All fig in	degrees			2500BC	
6th May	16	63	297	Beltane	31st May	16.6
21st Jun	23.5	51	306	Mid-Summer	13th July	23.9
5th Aug	15.8	63	297	Lughnasa	30th Aug	16.7
23rd Sept	0.3	90	270	Equinox	9th Oct	0.4
6th Nov	-16.3	115	245	Samhian	22nd Nov	-16.6
21st Dec	-23.5	129	231	Mid-Winter	13th Jan	-23.9
5th Feb	-16.1	115	245	Imbolc	28th Feb	-16.7
21st Mar	0.3	90	270	Equinox	9th April	0.4

Cross quarter day - Table adjusted and extended from Aubrey Burl's table (1990) for 50° location.

The cross-quarter dates fall currently between the 2nd and the 7th every third month. The name of the date from the current Julian calendar changes, therefore it is important to understand the declination of the Sun and its rising and setting points. In practice pinpointing the position at cross quarter would have proved difficult and cross quarter alignments have been heavily criticised due to the difficulty and ambiguities involved.

The Moon

The Moon rises in the east and sets in the west in the same manner as the Sun. Over the course of a lunar month the Moon goes through a number of phases. It takes 29.5 days for the Moon to appear to us as the same phase in the sky, or for instance, the time it takes from one full Moon to the next full Moon. In the northern hemisphere the full Moon rises in the east around 6pm in the evening and sets in the west around 6am the following morning. Understanding which phase the Moon is at and the time of day means that you should know where to find it in the sky. It is worth considering these phases and what time and where we can see the Moon.

After the full Moon, the Moon starts to wane. It rises later in the evening after Sunset, sometime glowing red when it's near the eastern horizon, it sets in the west after Sunrise. As the Moon moves through the days it rises later each evening, with the third quarter Moon rising around midnight in the east and setting around midday in the west. Eventually a crescent Moon can be seen before dawn in the east. This waning crescent Moon is also known as an old Moon. The waning crescent recedes leaving just a small strip on the left of the Moons disc, visible in east in the dawn light. This can be seen in the image taken on summer solstice Sunrise 2017 at Tregeseal Stone circle, looking east.

Waning crescent Moon over Tregeseal circle in the dawn light 21st June 2017

The following day no Moon will be seen and the Moon is called "new". The next night just a small crescent of the Moon will appear in the west setting just after dusk. The crescent will be to the right of the Moons disc and known as a waxing crescent Moon. The Moon will then continue through the phases with a first quarter Moon rising around midday and setting in the west at midnight. By the time it reaches waxing gibbous it is rising in the early afternoon and sets in the west in the early hours.

The Moon takes 29.5 days to move through these phases, but it takes 27.3 days to make one orbit of the Earth (known as is sidereal month). This is because for the Moon to reach the same orbit position with the Earth and the Sun it takes another 2.2 days. That is why the phases take 29.5 days or 2.2 days longer than its orbit. Understanding the motion of the Moon complex. But it doesn't mean that people haven't observed the Moon and understood how its motion changes through the years. People have understood that it changes speed along its path against the background of the stars since Babylonian times. Certain patterns could have been observed and predictions on the motion of the Moon could have been made even if the mechanics weren't fully understood.

We know now for instance that if we were to observe the Moon at the meridian one night, after one sidereal month it would be found within the same position of stars but not the same place in the sky. It would take nearly three sidereal months to be in the same position for the observer. But it will be about one hour east of the meridian after three sidereal months and at a different phase. Every three sidereal months the lunar phase will slip by 0.234 lunation, as calculated below.

$$\Delta\varphi = \frac{88.5177\,days - 81.6950\,days}{29.5059 days} = 0.234\ lunation$$

The regression of the lunar nodes has the consequence that the range in declination changes over time. This means that as the Moon moves along the horizon with a motion of +/-ε each day, for each lunar month the most northern and southern extremes move. The pattern this made along the horizon could be observed, even though the monthly difference in the motion would have probably been too small for them to measure and the pattern would take 18.61 years for each cycle to complete, as it takes the Moon 18.61 years to reach its extremes of declination.

These extremes are known as major and minor lunar standstills although it is worth pointing out that the Moon does not standstill and these extreme points may not be visible as per the location of the Moon in its cycle. So how likely would they have been able to observe and measure this?

The extremes of this cycle would have been particularly observable at higher latitudes. For instance, in Scotland when the Moon reaches the lunar major northern standstill position it can almost appear circumpolar skimming the horizon and in the right weather conditions this is an event which would happen every 19 years and could have become part of a collective memory. But even in Cornwall this cycle could have been observed. It is quite a simple task to do, If a person observes the rising or setting full Moon closest to the winter solstice. They will observe a Moon which is rising at its farthest azimuth to the north. (Likewise, the full Moon closest to the summer solstice would rise at its most southernly azimuth through the year). All it would take is for someone to measure and mark this position annually to see that this position moves year upon year. It would take years for the Moon to rise in exactly the same position as it did in the first year. They would find after 19 years that they had marked the extremes of the lunar cycle having observed this 19-year cycle in full.

How do we calculate the lunar major and minor declinations for Cornwall?

The diagram below shows how this motion occurs with the monthly setting (or rising) positions of the Moon along the horizon increasing to the lunar major positions and then decreasing again towards the lunar minor positions. At lunar major the Moon with rise and set in the most northerly and the southerly positions possible, with the widest range of declinations along the horizon during the course of a month. The opposite is true for the lunar minor part of the cycle.

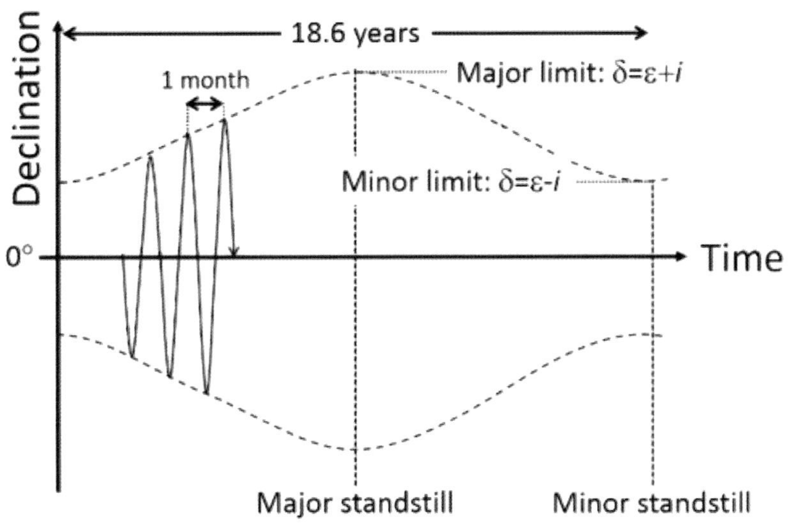

Motion of the Moon over 18.61-year period.
Credit Clive Ruggles (2016) Astronomy in Prehistoric Britain and Ireland

The major and minor standstills have a value of

$$\delta = \varepsilon = / - i$$

The extreme of the lunar node being i.
The current value of i is 5.1°.
Lunar parallax also must be considered in all figures need to be reduced by a value of P = 0.85

Let's consider 2500BCE When ε = 23.9
The major standstill declinations are 28.15° and -29.85°.
The minor standstills declination is 17.95° and -19.65°.
If viewing the lunar lower or upper limb these figures need to be adjusted by 0.25°

We can now consider how the Moon may have been observed and why
There are no documented lunar standstill examples in Cornwall, Merrivale in Devon is the closest example. But, they are worth consideration due to the myths which surround the circles; particularly Boscawen-ûn having a lunar connection. These myths include links with the quartz stone at Boscawen-ûn, the number of stones in the circles being 19 and local cup marked stones. Links to major and minor lunar positions have been made with cup marks in stones. The Merry Maidens has a heavily cup marked barrow located near it called Tregiffian. At Boscawen-ûn cup marked stones are located to the north at Goldherring and Leah.
It is worth noting that the full Moon closest to summer solstice is at its annual minimum declination and will culminate at its lowest altitude. The reverse happens at the winter solstice with the full Moon nearest to this date at its highest declination rising to its highest altitude in the sky. In 2500BC the 99.8% illuminated Moon closest to the winter solstice

would have reached a visually impressive highest southern altitude of 67°44'06''. Whereas the full Moon closest to the summer solstice would reach a maximum southern altitude of only 11°10'27''.

The key solar and lunar declinations

Taking the main three solar alignments and the lunar major and minor extremes gives seven rising and seven setting positions on the horizon. These are the main declinations which will be considered at the stone circle sites. They can be visualised in the following manner.

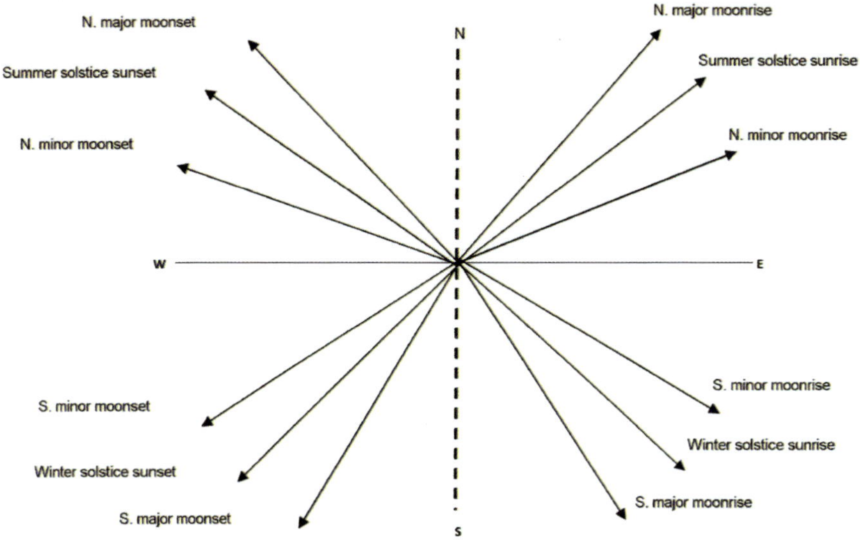

The respective declinations (in degrees) for these 14 positions that will be considered for alignment with horizon targets are shown in the table below

Sunrise	2500BC
Major North	28.15
Solstice summer	23.9
Minor North	17.95
Equinox	0.4
Minor South	-19.65
Solstice winter	-23.9
Major South	-29.85
Sunset	
Major South	-29.85
Solstice winter	-23.9
Minor South	-19.65
Equinox	0.4
Minor North	17.95
Solstice summer	23.9
Major North	28.15

Stellar alignments – what is the problem?

To answer this question, we first have to understand a little about the motion of stars and how the position of a star changes with time.

All stars in the northern hemisphere proceed around the NCP in a circular motion clockwise. The NCP is also in the direction of the cardinal point of north from the site. Stars at Boscawen-ûn with a declination of 40° to 90° are circumpolar. Stars with declinations down to 39°54'37'' would rise above the horizon over the course of the year. Even with perfect seeing conditions people could only see naked eye objects in the night sky up to and inclusive of apparent magnitude 6. Only the brightest of stars could be seen near the horizon due to atmospheric scattering and haze and with extinction, a star only becomes visible when its altitude equals its magnitude.

Polaris is a 1.97 magnitude star, it currently has a declination of +89°15'50'' and is positioned currently near the NCP. Due to precession, this hasn't always been the case. Precession occurs due to the gravitational pull of the Sun on the Earth and makes the NCP change its position over a period of 26000 years. The change in the NCP can be seen in figure 36. In 3000BCE a naked eye star called Thuban would have been close to the NCP. The next naked eye star at the NCP was Kochab between 1900 to 1100BCE.

Stellar alignments would be difficult to prove considering the lack of an accurate date for the building of circles such as Boscawen-ûn. The change in declination due to proper motion, as well as precession, makes much of the sky covered by alignment opportunities. The table below shows the brightest 17 stars position over time, also included is the open cluster the Pleiades as per the suggested alignments of Lockyer. This data has been reconstructed using Stellarium with an error of approx. 0.5° at 2000BCE. For example, Arcturus a bright circumpolar star has a changing declination from 48°33'57'' in 3000BCE to 40°22'44'' which is a range in declinations of 7°46'15'' over 5000 years.

Star	App	Dec	Dec2	Dec3	Dec4	Dec5	RA
Name	Mag	3000BC	2500BC	2000BC	1600**BC**	J2000	J2000
Sirius	-1.46	-22°21'15"	-20°42'47"	-19°16'55"	-18°17'55"	-15°25'28	6hr48m22s
Arcturus	-0.04	48°33'57"	45°44'31"	42°47'42"	40°22'44"	21°24'54"	14hr21m0s
Vega	0.03	43°59'58"	42°36'06"	41°23'05"	40°32'46"	38°26'25"	18hr36m20s
Capella	0.08	26°10'08"	28°53'10	31°33'38"	33°38'33"	46°26'38"	5hr15m40s
Rigel	0.12	-25°37'54"	-23°04'02"	-20°37'32"	-18°46'39"	-8°12'04"	5hr14m31s
Procyon	0.38	2°58'48"	4°31'55"	5°49'32"	6°39'36"	6°15'46"	7hr42m56s
Betelgeuse	0.5	-7°27'02"	-4°57'60"	-2°38'03"	-0°53'50	7°23'45"	5hr55m1s
Altair	0.77	9°58'52"	8°34'10"	7°25'26"	6°42'35"	8°28'33"	19hr48m0s
Aldebaran	0.85	-4°46'48"	-2°02'23"	0°39'18"	2°45'17"	16°42'25"	4hr35m35s
Antares	0.96	-4°08'35"	-6°55'16"	-9°40'28"	-11°50'08"	-26°24'23"	16h29m26s
Spica	0.98	14°42'23"	12°35'39"	10°16'60"	8°18'55"	-11°07'23"	13hr25m24s
Pollux	1.14	22°51'10"	24°49'07"	26°31'51"	27°41'40"	28°11'30	7hr48m56s
Fomalhaut	1.16	-43°58'06"	-44°11'06"	-43°57'28"	-43°27'45"	-29°25'25"	22hr55m50s
Deneb	1.25	36°30'34"	36°28'47"	36°39'38"	36°57'23"	45°16'43"	20h41m25s
Regulus	1.35	23°36'39"	23°55'41"	23°52'48"	23°34'45"	11°54'50"	10hr09m41s
Adhara	1.5	-33°38'01"	-32°09'34"	-30°53'29"	-30°01'54"	-28°58'24"	6h58m36s
Castor	1.57	24°51'48"	27°00'26"	28°54'51"	30°14'31"	32°04'45"	7hr35m40s
Pleiades	1.6	0°05'22"	2°49'59"	5°36'07"	7°48'42"	24°22'03"	3hr45m49s

Brightest stars at latitude 50° and their changing declinations.

"Indeed, the inclusion of stellar possibilities may actually hinder rather than assist those aiming to demonstrate the broad astronomical potential of a site or set of sites, since they are inherently more numerous and flexible and tend to engender scepticism among those who are well aware of the methodological and data selection issues." Clive Ruggles

Another consideration is that a star is a small point of light. Many of the monuments in Cornwall are built from large wide topped pieces of granite, which makes making a stellar alignment even more complex. Additions of alignment holes or slits may have been a possibility to make the alignment more accurate in practice, and these will be considered when looking at the individual circles and stones. Stellar alignments were suggested on numerous occasions by historic archaeoastronomers. For instance, John North considers that the earliest of sites were dedicated alone for the observation of the stars. Historic suggestions for stellar ideas will be examined with caution. This will include suggestions by Lockyer of alignments with Capella, the Pleiades and the circle.

Heliacal rising and setting stars

Due to the tilt of the Earth, stars are seasonal. So which stars are visible in the night sky depends on the time of year. For instance, the constellation is known for being visible in the winter, but not necessarily in the summer. This has raised the questions; What is the first date that an astronomer can see a particular star rise? What is the last date they can see a star set before it disappears from view for the season?

These times are known as the helical rising and helical setting dates of a star. The rising star will be rising in the dawn light before the Sun rises bringing daylight, heralding the first time that star has returned for the season. The setting star will be the last time it is viewed setting in the west just after Sunset, disappearing from view for a number of months.

The following table gives examples of these dates, at a latitude of 50°. If you are observing a star at the horizon these dates and times are approximate as there will be local seeing conditions which would change on a daily basis. An overriding rule is that a star at the latitudes of Cornwall will need to be at least 7° in altitude difference to the rising Sun, which is still below the horizon. The star will need to be at least 10° altitude difference, at the horizon, to the Sun when it is in a setting position. The amount of distance the star will have to be away from the Sun depends on the stars magnitude, with brighter stars such as Sirius becoming visible closer to the rising or setting Sun than a dimmer stars.

Sirius 3 °, Sun -7 ° / -10 °			
m = -1.46	date	time of day	azimuth
Heliacal Rising	26th Aug	05:48	122 °
Heliacal Setting	1st May	21:44	237 °
Arcturus 3 °, Sun -7 ° / -10 °			
m = -0.04	date	time of day	azimuth
Heliacal Rising	10th Oct	07:05	62 °
Heliacal Setting	5th Dec	16:11	298 °
Rigel 4 °, Sun -8 ° / -12 °			
m = 0:12	date	time of day	azimuth
Heliacal Setting	22nd April	20:33	251 °
Heliacal Rising	4th Aug	05:03	109 °
Procyon 4 °, Sun -8 ° / -12 °			
m = 0:38	date	time of day	azimuth
Heliacal Setting	May 29th	21:51	273 °
Heliacal Rising	1st Aug	05:27	87 °
Betelgeuse 4 °, Sun -8 ° / -12 °			
m = 0.5	date	time of day	azimuth
Heliacal Setting	13th May	21:14	277 °
Heliacal Rising	1st Jul	04:49	83 °
Altair 4 °, Sun -8 ° / -12 °			
m = 0.77	date	time of day	azimuth
Heliacal Rising	26th Dec	07:44	81 °
Heliacal Setting	30th Jan	17:35	279 °

Other alignments planetary /events

Five planets would have been visible to the naked eye. These are Mercury, Venus, Mars, Jupiter and Saturn. The motion of these planets is measurable but at times could be considered as complex. Out of all these planets Venus is worth most consideration for alignments as it's the brightest planet making it easily identifiable.

Venus above Boscawen-ûn (2017)

Venus has a relatively short period, meaning that patterns can be observed forming when considering its motion. Venus current inclination of 3°24' and a period of 225 days. Its synodic period or when it reaches the same configuration with the Earth is 583.92 days. For the first 268 days of this synodic period it appears in the night sky as a evening star, setting in the west shortly after the setting Sun. It then hides in the glare of the Sun for 50 days, before becoming a morning star rising before the Sun for 258 days, finally hiding in the glare of the Sun for the last 8 days of the synodic period.

Venus has extremes to its rising and setting points and these can be calculated in the following manner;

$+/-(\varepsilon + v)$

Where v = inclination of Venus 3.4°

If ε = 23.9 (as per 2500BCE)

+/-(23.9+03.4)
Extremes of Venus in its cycle = +/- 27.3°

To calculate the duration of the cycle we need to look at Venus synodic period, or the amount of time that it takes Venus to return to the same place if viewing it from Earth. This is 584 days. This can be compared to the amount of days in the Earths year.

We can see that
584 x 5 ~ 365 x 8
These extremes of the cycle of Venus happen every 8 years.

Mercury's extremes can be calculated in the following manner, although it would have been more difficult to observe due to its location in the sky nearer Sun.
Mercury's extremes are calculated as follows

+/-(ε + m)

where m = Mercury's inclination = 7°

If ε = 23.9 (as per 2500BCE)

+/-(23.9 + 7°)

Extremes of Mercury in its cycle = +/- 30.9°

To calculate the duration of Mercury's cycle we need to use its synodic period, which is 116 days. This shows that the cycle is short and only lasts 1/3 of our year.

Mercury is a lot more difficult to observe due to its location next to the Sun and would be a lot less likely to be used in this manner.

If we now consider the major planets Mars, Jupiter and Saturn. They are not as bright as Venus. But their motion would be observed and possibly understood. Retrograde motion was understood by the Babylonians and ancient Greeks who used this to make predictions of the future. It is worth explaining what this motion is, but it is unlikely that any conclusive evidence will be found to show that retrograde motion was built into the circles design.

The major planets have an apparent motion where they progress nightly with a forward motion through the night sky. This means if you were to go and sit and look at Mars one night and at the exact same time the second night it will have moved further east in the night sky. This movement happens as it rises in the east earlier each evening. Occasionally this motion is reversed and following on is a shorter period where the motion of the planet appears further west each night when observed at the same time. It appears as though it is moving in the opposite direction, this is known as retrograde motion.

When we examine the motion of Mars, retrograde motion occurs within a complete period of 25.6 synodic months lasting 72 days in duration. If Mars is visible in the morning sky Earth is approaching it within their respective orbits. When Mars rises at Sunset opposition of the planet is approaching, finally if Mars is already arisen by early evening dusk it has past opposition.
Jupiter has retrograde motion events occurring every within every 13.1 synodic months and they last 121 days. Saturn's occur with every 12.4 synodic months and last 138 days.

Eclipses, Aurora, Comets and meteor showers.

The most fabulous natural events can happen in the sky. They are awe inspiring, often unpredictable and often held a place of fear to the unprepared.
Alignments to events such as eclipses are more complex and will not be considered in this work.

Other important considerations
Refraction, Parallax and extinction

The effects of refraction are most pronounced when looking at a celestial object at the horizon, the effects decrease with increases in altitude. The amount of refraction varies due to temperature and air pressure. A figure for refraction at the horizon is used as 32' (0.54°). When the Sun or Moon is first seen it is still a whole disc amount below the horizon. It means that we see the Sun about two minutes before Sunrise and around two minutes longer after it sets, as shown on in figure 28. This amount should be considered as variable as this value changes with altitude. By an altitude of 2° the light from objects will be refracted by an amount of 0.3°. A maximum horizon value of refraction of 0.54° will be added to the error.
The light from stars and planets is also affected by the larger amount of atmospheric airmass at the horizon. Following is an approximation that can be used when observing stars with the naked eye up to magnitude 6. A magnitude 1 star will only become visible at around 1° above the horizon and a magnitude 2 about 2° and so on until magnitude 6. This is photographed below, where magnitude 2 stars in Lepus are difficult to see compared to stars of a similar magnitude in Orion

Extinction of Starlight (2017)

Parallax has a real and dramatic effect on the position of the Moon. Lunar parallax will be considered as 0.85°. Stellar Parallax measured against the semi major axis of the Earth orbit is less than 1 arc-second and will not be considered for this project.

The combined error for this project will be 1.04°

(Lunar parallax will be considered within the lunar calculations)

Boscawen-ûn

Boscawen-ûn stone circle is an evocative site. Nestled to the south of Creeg Tol a local granite outcrop, it is in a liminal location. You can approach the circle along footpaths from two main directions. The first footpath lays to the north and parking is in a lay-by on the A30, the second access is along a public footpath which runs by the farm in the east. Both approaches give you a sense of the remoteness and a rural tranquil positioning of the circle. It is a secluded site with many people driving past on their way to Landsend via the A30 above, unaware of the wonderful treasure of a circle site close-by. The first time I approached the site from the north. In this direction are the remains of a local bronze age village Goldherring and you can imagine people from the past making the same journey down the hill to the stone circle. Part way along the path it is worth taking a small detour. Creeg Tol is a granite outcrop which gives commanding views of the southern landscape below. It is also from here that you will get the first view of the circle itself. The solitary circle stands nestled in the dip in the landscape, but far from feeling abandoned or alone the site gives a sense of being cherished, having existed here across the millennia in its current form.

In many ways, during the past year, Boscawen-ún became a second home to me. While waiting for Sunrises and Sunsets I have observed the change in the seasons at the circle. But one thing remained the same and that was the tranquillity of the site, Often, I was alone in the circle sometimes for hours on end. One of these visits comes to the forefront of my mind. Having risen when it was still dark, I drove to the circle with the beginnings of dawn, hoping the low cloud on the horizon would clear. I arrived in time for the Sunrise of the 25th June 2016, the week had been wet, and the solstice had passed behind a thick blanket of cloud. I stood atop Creeg Tol willing the low bank of cloud to blow out of the way, even though I knew I was days late to see the summer solstice Sunrise. The vantage point of Creeg Tol meant that I would see the Sun peer above the horizon, something that I could not replicate in person in the circle below due to the large hedge obscuring this direction.

The dawn had a stillness about it which makes it one of my favourite times of the day. The clouds were starting to disperse and right on schedule, the Sun started to peer above the horizon. I photographed the sunrise from my vantage point at Creeg Tol and set off down the hill towards the circle. About half way down into the hill I started to lose the Sunrise, the Sun was setting behind the hill it had just risen from. By the time I reached the circle the Sun was once again well below the ridge on the horizon. I realised that without the hedgerow I could witness the Sun rising twice - a double Sunrise. Once from atop of Creeg Tol and then again from inside the circle. I hoped this would work in reverse: with the Sun setting visually from the circle and once again from Creeg Tol. It was an idea I would test out repeatedly over the summer months with great success. I think this double Sunset and Sunrise during the summer months is one of the most visually beautiful aspects about the circle. The local bronze age settlement Goldherring was to the north of the site with access to the circle from the direction of Creeg Tol. Double Sunsets and double Sunrises are something that we can all witness from the site and this is only the beginning of what makes Boscawen-ún astronomically special.

The circle has 19 stones, with a central leaning stone in the middle. It stands in an ovoid shape with a gap for an entrance or extra stone. The stones in the image below have been numbered, with number one being the first stone after the gap, running clockwise around the circle.

The circle isn't large - it has a diameter of 24.9 by 21.9m, the stones themselves are fairly short and stand between 1.30m and 0.90m high. They are in the main rounded on the head with flattened sides facing the centre of the circle. All are granite except for one stone which is made from a large block of quartz. The quartz stone to the southwest of the circle, numbered 18 in the image above. There are also remains of a burial cist located to the northeast of the circle. The stones which form the cist are numbered 7 and 8. There are more stones which form this feature, but they are laying down either side of the circle and are not numbered. Boscawen-ûn circle is well preserved. The earliest record of the site by William Camden in 1586 wrote that "In a place called Biscaw Woune are nineteen stones in a circle, twelve feet from each other, and in the circle, stands one much larger than the rest."

Only three stones have ever been recorded as being replaced and these are stone numbers 12, recorded by (Borlase 1754 p209) and possibly stones 5 and 9 from Cotton's plan 1827 which can be seen below. Edwin Dunkin a Cornishman and astronomer paid a visit in 1870, he describes how there are 18 stones still standing and one fallen. He considers the site in good repair at that time. What he fortunately did not see was the hedge which had cut though the centre of the site and had been removed by concerned antiquarians during the decade before.

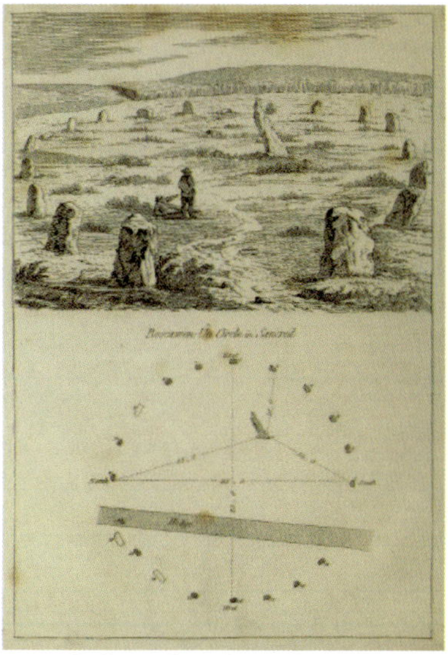

Drawing and plan of circle William Cotton (1827)

The hedge can be seen in Blight 1861 drawing (below) and also Cottons plan above. This hedgerow was removed by 1865, at the same time three fallen stones were re-erected. It was also during this time that the site was encased by the existing hedgerow which surrounds the circle.

BOSCAWEN-ÛN CIRCLE.

The impressive and unusual central leaning stone would have stood 2.4m high if it had not been in a leaning position. It currently leans at 45+/-2° and early excavations in 1877 suggested that this stone was placed in this manner and was never upright. William Stukeley drew the circle in 1749 and from his diagram it is difficult to see if the stone is leaning, certainly five years later when local antiquarian William Borlase visited the site he found the stone in a leaning position.

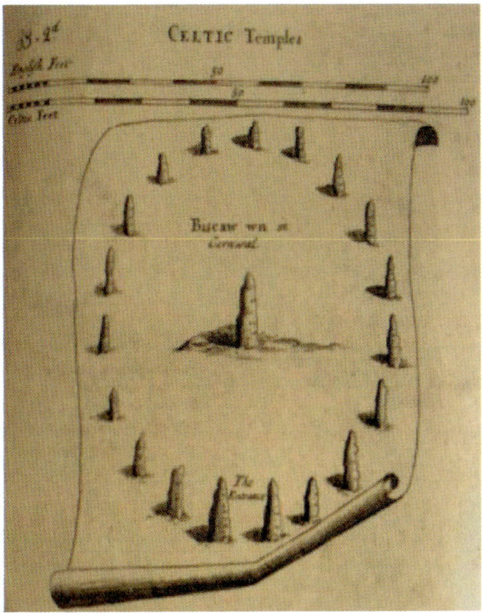

William Stukeley 1749 Intinerarium Curiosum

The eighteenth-century drawing (shown below) by William Borlase showing the central stone with an exaggerated lean up to 60°.

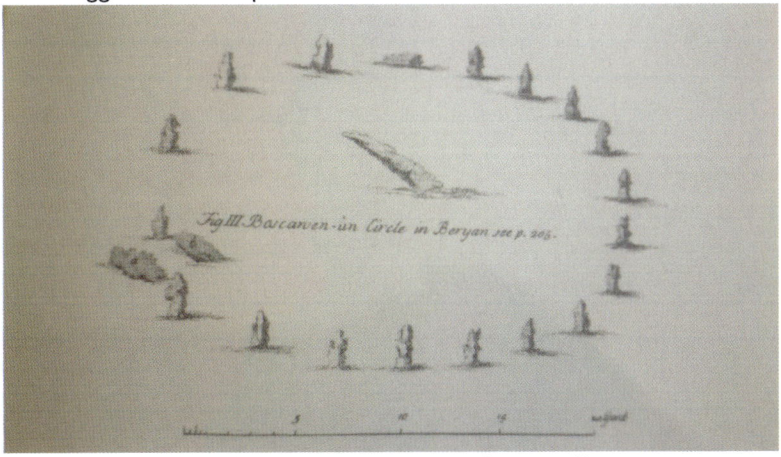

Borlase 1754 Antiquities of Cornwall

It has been claimed that the central leaning stone resembles a large axe head in shape. Axe shaped monuments are prevalent from this time. Other examples of this can be seen in Brittany, just across the channel to Cornwall, where axes are given their most ritual role and were considered to be thunder axes with a possible astronomical connection with either the Sun or with Jupiter. (Kelley 2011).

It was only recently that rock art was discovered on the north side of the central stone. This was in 1986 by local antiquarian and author Ian McNeil Cooke. It was originally thought to depict axe heads, but 3D imaging by Tom Goskar showed it to be possible feet and circles or breasts. A photograph of the art can be seen below.

Rock art depicting feet is uncommon and only occurs at a few other locations in the UK including the Caulderstones in Liverpool and the Priddy cist in Somerset. Herring (2014) suggests that the art has the most similarity to the feet discovered at Dolmen du-Petit-Mont at Arzon, Brittany. Two spherical shapes can be seen above the feet in the 3D image. The one to the left has spiral lines coming from it. It has been suggested they are breasts (Herring 2014) but could possibly depict the Sun and Moon.

Boscawen-ûn in the local landscape

The circle sits in a liminal location, between the lowland south and highland north. It lies in a dip in the landscape and the north side dominated by Creeg Tol a local granite outcrop. Neolithic archaeological features found on Creeg Tol include barrows and menhirs.

Perhaps the most distinct horizon feature when stood in the circle is that of Chapel Carn Brea. Lying to northwest Chapel Carn Brea is beautifully framed by the local the local hills of Creeg Tol and the Bunkers Hill ridge. The framing of this hill is very distinctive when standing within the stone circle and is perhaps to most immediately visual horizon feature at Boscawen-ûn. It is interesting to note that this view would be lost if the circle was moved a short distance in any direction. Chapel Carn Brea is considered a sacred site since prehistoric times with a large number of barrows located on it. These include the Scillonian barrow which is easily visible on its summit even from the distant view when stood in the circle. Looking southeast there is a long ridge broken by two sea glimpses. One in a small dip in the horizon and then further east the Lamorna Gap. The Lamorna Gap is one of the few access points to the sea in West Penwith and would have been considered a vitally important location in the landscape.

Inter-visibility over the southern landscape would have been good. Even today the distant Merry Maiden's circle complex is visible. Peter Herring in his 2014 archaeological assessment of the site suggests a processional route from the Lamorna cove via the Merry Maidens extending to Boscawen-ûn and beyond.

Myths, Legends and Traditions

There are many stories worth telling about the Boscawen-ûn. The site has numerous traditions and legends. First is a myth about the site, which is very similar to the tales told about the nearby Merry Maidens. Secondly, there is a piece about the Bardic Cornish Gorsedd in 1927. This event was held here for a number of years, due to a proposed (but now questioned) link to a 6th century Welsh Triad. Finally, a piece about the sites link to the Anthurium Legend.

A local myth - Taken from Western Morning news 8th August 1893

"Legend said once upon time nineteen Merry Maidens were dancing there Sunday, wicked girls, to music furnished by piper and two fiddlers. piper stood in the middle of the circle and the fiddlers outside. They scraped and squeaked out some old world dance tune, and the feet

kept time merrily. happened that Perseus who had just cut off the head Medusa was flying slowly and wearily to his homo in Sunny Greece, born aloft the winged sandals and made invisible by the cap of darkness. His righteous soul was vexed at the sight of those Sabbath-breakers and dropping down into their middle drew from under his cloak that sad and awful head, holding it up to the view of all. Rooted to the spot the dancers crystallised into granite stones, but one who had had qualms of conscience became snow-white quartz. The piper also tamed to stone, and the two fiddlers ran away, but were caught few hundred yards off."
The tradition of the Gorsedd at Boscawen-ûn.

Taken from "The Cornishman" 27th June 1927. It discusses a visit to the circle by the 'Old Cornwall Society'.

"BOSCAWEN-ÛN CIRCLE. The sky was heavily overcast, and visibility was poor, but nevertheless a pleasant run was made along the Land's End road to Boscawen-ûn, where the first halt was made. The cars had stop outside the entrance, because the lane was too narrow take them, and the party made their way, on foot, over the somewhat rough a mile of lane, which terminated abruptly the Boscawen-Us circle. The circle is about ninety feet in diameter, and stands out rather forlornly from piece of moorland. Its perimeter formed nineteen large stones, whilst the centre is marked a still larger stone which stands out alone. Here party reassembled and awaited the appearance Mr. Henry Jenner. In a few minutes that gentleman arrived, attired his Bardic robes, and a splendid figure presented. His big frame, his expressive face and white beard lent a dignity to his robes, and with very little mental effort, one might easily have imagined that one was back in the days of the ancient Druids and bad come to attend a Gorsedd. Referring to the stones, Mr. said the circle of Boscawen-ûn was the most noteworthy stone circle Cornwall, but before told them why, he would like explain that it was the practice of the Welsh Bards, in quite early times, to tabulate their sayings in triads, or sets of three. ....There was one Welsh triad which interested them very much that day, which mentioned the three principal session places of the Bards is Britain, i.e., the session place of the Hill of Meriw, the session place of Beisgawen, and the session place of the Hill of Gwyddon. A variant this triad gave Beisgawen as Beisgawen in Dyfnwal, or Damnonia. The intention seemed to locate one in each of the principal divisions of Britain. They were standing in in Beisgawen of Dyfnwal at that moment........He did not know when a Goreedd was last held at Beisgawen, or what sort of affair it was, but the circle was probably not originally built for such purpose, though it had not yet been decided by antiquarians why such circles were constructed. There were many theories, but the most probable seemed to that they were temples with a special orientation, which might indicate the seasons through the rising of certain stars. One thing seemed certain, however, and that was that these circles and their attendant menhirs, belonged to the new Stone Age, perhaps about 1800 B.C."

The Legend of Arthur also taken from the same article.

"One other item of interest about the stones was their name. There were three places called in Buryan parish. Boscawen-ûn, Boscawen on the Down, and Boscawen Rose. It was commonly held that Boscawen meant the dwelling of elder tree, but thought this was a mistake. The epithet in beginning with "bos" was almost invariably a personal name, and he was more inclined to interpret Bosgawen as "the dwelling of Gawain." Whether it was called

after Gawain, King Arthur's nephew, he could not say; but close to Boscawen Rose was "Rosemodres," which could only mean "the heath of Modred," who was another nephew Arthur. There was a definate tradition of battle of Arthur on Vellandruchar Downs, on the river Kemyel which flowed into Lamorna Cove. If Kemyel is Camlan, where Arthur and Modred killed each other, Gawain having died not long before the battle, they had some apparent reason for these names"

Historic archaeoastronomy measurements at Boscawen-ûn

There have been several historic astronomical links made with Boscawen-ûn. These were suggested since the mid-19th century and by a number of professionals both from an astronomical and archaeological background. It is worth looking at these in chronological order.

The earliest suggestion that Boscawen-ûn had an astronomical link was made by Richard Edmond (1862). He suggested that all four circles within West Penwith had 19 stones and that they may have been designed to consider the Metonic cycle.

What is the Metonic cycle?

The Metonic cycle shows that 19 solar years and 235 synodic lunar months (phases) are very close, they both are nearly 6940 days. There are only 2 hours difference between the two cycles.

Edwin Dunkin (1870) discusses Edmonds idea and casts a small amount of doubt onto it, he lists several texts showing that anywhere between 18 and 20 stones could have been originally positioned here in a circle. His considered ambiguities lie in whether authors count the central stone as part of the 19, or whether the gap had a stone within it. He claims that just eight years after Edmond's visit there are only 18 stones in the circle but his diagram (Figure 17) shows 19 and a central stone. He concludes that he considers the circle to have 19 stones and although an astronomer he offers no astronomical alignments for this circle within his paper.

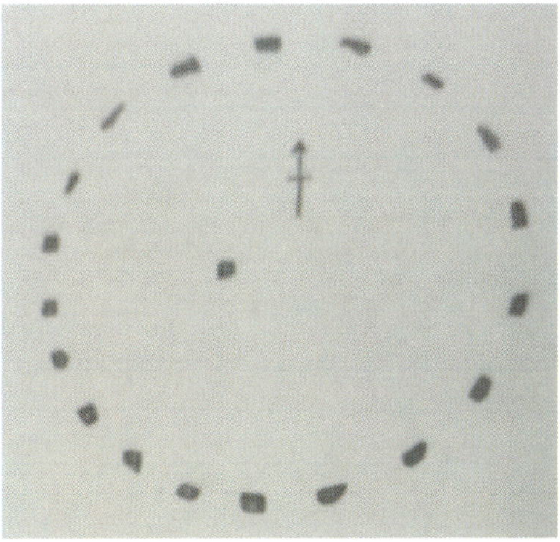

Dunkin's diagram of Boscawen-ûn. – Remarks on the Stone Circles of West Cornwall Edwin Dunkin, Reliquary Quarterly 10, 1869-70.97-102

The main archaeoastronomy work at Boscawen-ûn was undertaken by Norman Lockyer in 1905. The weather was poor the day he visited the site the and he made these predictions from a map. He would make seven suggestions for the site and it is worth looking at these individually. The map and the table with Norman Lockyer's alignments is shown below.

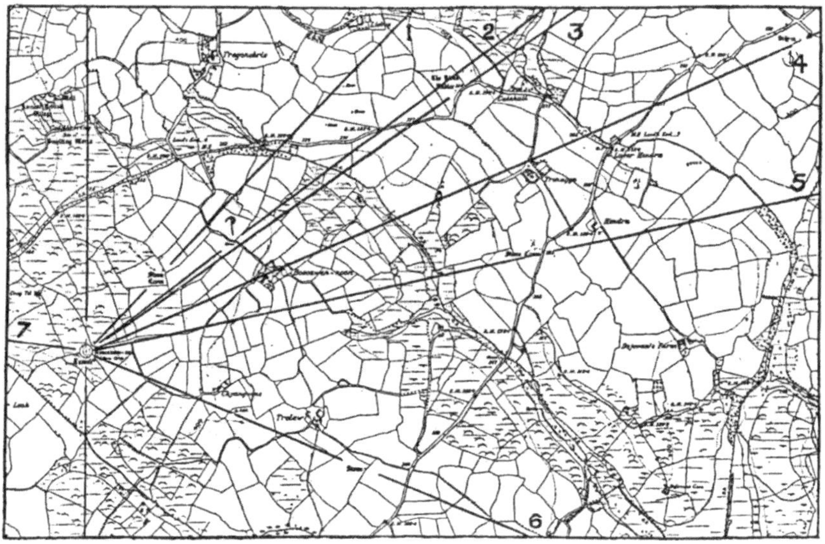

Marks.	Az.	Hills.	Dec.	Star.	Date.
1. F. Stone cross	N. 43° 15′ E.	2° 7′	+29° 26′	Capella	2250
2. P. Fine menhir	N. 53 30 E.	1 15	22 58	Solstitial sun	—
3. B. Blind Fiddler	N. 54 30 E.	1 15	22 24	..	—
4. Two largo menhirs	N. 66 50 E.	1 0	14 55	May sun	—
5. Stone cross	N. 78 0 E.	1 0 (?)	+8 8	Pleiades (May)	1480
6. Stone	S. 66 30 E.	1 0 (?)	14 32	November sun	—
7. Stone	N. 83 30 W.	1 0 (?)	+4 36	Pleiades (September)	2120

Credit Norman Lockyer (1906) Stonehenge and Other British Stone Monuments Astronomically Considered.

Lockyer made four predictions involving the Sun. Two for the solstice and two for Sunrise on cross quarter days (midpoint between solstice and equinox). The final three predictions were involving stellar targets, two rising and one setting. Many of the alignments were to objects which are not visible from Boscawen-ûn. Following are these suggested alignments in more detail.

Alignment 1 - Lockyer's stone cross is the Boscawen Un field menhir. He suggests that the star Capella would be visible rising late spring early morning 2250BCE. Lockyer gives Capella an azimuth of azimuth 43°15 and declination of +29°26′. This is very close to the results from modern planetarium software which gives an azimuth of 42°13′19″ and declination of 30°13′40. Lockyer offers a date for when Capella will be visible at this stone. He does this as the rising and setting positions of stars change over time. They do this due to the changing obliquity of the ecliptic and the fact that the Earth's equator changes its angle to the plane of the Earth's orbit. The stars in the sky change their position with the change in the Earths tilt on a 26000-year cycle. This means that if you have an accurate date of a monument you can suggest stellar alignments but Lockyers attempts to do this the other way around and suggest dates to monuments from alignments to stars. Much of the criticism levelled at Lockyer stems from the fact that he tried to date sites from stars rising positions.

Alignments 2 and 3 - Summer Solstice Sunrise. Lockyer suggested two menhirs as alignments for the solstice rising position. Neither are visible from the stone circle. The first is the Boscawen-ûn hedge menhir at 625meters distance. He suggests an azimuth of 53°30 and a declination of 22°58′ for 2500BCE. Planetarium software gives the rising solstice Sun in 2500BCE at azimuth 52°45′47″ at a declination of 23°46′10″. His measured azimuth and declinations are both slightly lower than is expected at this epoch.

Alignment 3 with the Blind Fiddler extends over1.5km, this alignment isn't credible as not only is it not visible but Lockyer's declination of the Sun at 22°24 is even lower than expected at this epoch therefore his azimuth reads too high.

Alignment 4 - Lockyer has made the Drift Pair the rising point of the May Sun. This pair of menhirs are not visible and are distant over 2.6km with many intervening hills between them. Although the May Sun rises at approximately 65° close to Lockyer 66°5 it is unlikely that these could have formed an alignment due to the distances involved.

Alignment 5 - This alignment leads from the circle to Trenugga cross at 1.5km and eventually to Tresvennack menhir at 3km. A local hill stands in the way of any visual contact between the stone circle, the cross and the menhirs. Lockyer suggests that these mark the direction of the rising Pleiades in May in 1480BCE. The Pleiades rise at an azimuth of 78° in 1480BCE. Due to lack of indivisibility between the circle and the menhir this alignment is not credible and reasons given with alignment 1.

Alignment 6 is with a line of stones starting with the Chyangwens hedge stone, continuing through the Trelew, Toldavas and Castallack stones. The Chyangwens is visible from the site. The others are visible from each other. Lockyer suggests this is at 66 S (or an azimuth of 114°). Planetarium software shows the November Sun would rising at just over 113° in 2500BCE. This is in the position of the November Sun.

Alignment 7 - Lockyer suggests an alignment to a stone at 277°. This stone could possibly be the now removed Bunkers Hill East menhir which was visible from the site. A map by Michell (1974) puts it at a much lower azimuth of 256.6°, but this could be inaccurate, as this stone is removed. It could mean Chapel Carn Brea although this is at an azimuth 282°. The Pleiades set in this direction in October. Lockyer calculated the Pleiades to set at azimuth 277° and declination of 4°36 for 2120 BCE. This is remarkably close to the azimuth of 275°53′50″ and declination of 4°44′54″.

Modern Suggestions after Lockyer

The Society for the Astronomical Study of Ancient Stone Monuments, Penzance (1906) Visited the site had tea and made no observations, exclaimed a wish to revisit in the following years.

Alexander Thom (1962) surveyed Boscawen-ûn and categorised it as a flat circle. He made no suggested astronomical alignments for this circle.

John Michells (1974) mapped a number of non-astronomical alignments between Boscawen-ûn and the surrounding stones. He expands Lockyer's suggestions with material on related menhirs.

John Barnatt (1982) claims there are no known astronomical alignments with menhirs at Boscawen-ûn. (Barnatt 1982 p71). He criticises Lockyer's alignments saying *"astronomical alignments claimed by Lockyer are not convincing; a tall menhir a short distance to the north-east was claimed to align with Capella in 2250BCE; other features are more distant and not visible "*. He continues his discussion saying that the quartz stone is orientated towards Candlemas[2], but he feels it would be difficult to make the measurements involved to prove this.

Aubury Burl (2001) states that Cornish circles have a tendency to have their largest stone at the WSW location, in the case of Boscawen Un this is a quartz stone. This aligns with the February and November Sunset.

Ian McNeil Cooke (1986) suggested through observation that the rock art he discovered on the central stone is aligned with the solstice Sun and the cist is located where the Sun rises on the solstice.

The results are all shown in the table below, they are on the whole inconclusive and methodology used by Lockyer has been extensively criticised. It was time that Boscawen-ûn was given a thorough survey and the alignments were tested with new onsite measurements which would draw conclusions to their validity or not.

Historic suggestions

Name	suggestion	Alignment	Measurement
Edmonds	19 stones	Metonic	
Lockyer	Boscawen-Un Field Menhir	Capella 2250BC	29°.43
Lockyer	Boscawen-Un Hedge Menhir	Summer Solstice sunrise	22°.97
Lockyer	Blind Fiddler	Summer Solstice sunrise	22°.4
Lockyer	Drift Pair	May Sun	14°.92
Lockyer	Trenugga Cross	Pleiades 1481	9°.13
Lockyer	Trelow line Menhirs	November Sun	14°.53
Lockyer	Stone Poss Bunkers hill Men?	Pleiades 2120	4°.6
Barnatt	Quartz stone	Candlemas	
Burl	Quartz Stone	Feb/Nov Cross Quarter	
Cooke	Central rock art	Solstice Sun sunrise	
Cooke	Cist	Summer Solstice sunrise	

Calculated alignments

After onsite measurements were made a comprehensive list of local horizon and manmade features was made. These of local sites included natural features such as sea views and Cairns, to manmade objects such as barrows and menhirs.

Following is the list of local features which are visible from the site, a number of local features have been discounted as they are not visible. This section will calculate the declinations of these features. Some of these would stand proud against the sky but many of them are only visible as objects within the landscape, they will be separated into H = horizon and L = within the landscape. The declination of all the objects were then calculated. Those sitting within the landscape will use the corresponding horizon altitude behind them. These figures are correct for someone standing in the centre of the circle and shown in table below.

---

[2] Candlemas is a Christian festival and falls at the start of February.

The declination is calculated using the following equation

$$Sin\delta = sin\lambda sinh + cos\lambda cosh cosA$$

Where λ = latitude, h = altitude and A = azimuth

All figures are given to an accuracy of +/-0.5°

Feature	Azimuth	Distance (m)	Altitude	Declination	
Boscawen Un field	41	364	1.5	28.9	H
Boscawen Un Barrow 4	93	105	0.4	-1.9	L
Chyangwens	110.7	716	0.3	-13.1	L
Trelow barrow 11	117	857	0.1	-16.9	L
Lamorna Gap	128.3	5000	-0.1	-23.4	H
Boscawen Un Barrow 2	132.5	36	0	-25.7	L
Piper I	133.2	3420	0	-26.1	L
Piper II	135.8	3430	0.1	-27.4	L
Pridden barrow 12	139	945	0.1	-29.0	L
Merry Maidens	140.9	3490	0.1	-29.9	L
Pridden	144.4	870	0.1	-31.4	L
Sea View	147	4830	0	-32.6	H
Higher Tevorian	154.9	1140	0	35.5	L
Pridden barrow 14	171	685	0.3	-39.3	L
Boscawen-ûn barrow 1	229-235	9	0.2	.-21.4 to 24.7	L
Boscawen Un Barrow 3	241	89	0.4	-18.1	L
Bunkers Hill 4	256.6	815	0.5	-8.6	H
Carn Brea	282	2680	1.8	7.7	H
Boscawen Un Barrow 6	310	409	2.3	24.3	H
Boscawen Un Barrow 5	317.8	260	2.8	28.3	H

The data is reduced to only those which stand against the horizon (marked H). These final seven targets shown in the table below.

Feature	Azimuth	Distance (m)	Altitude	Declination	
Boscawen Un field	41	364	1.5	28.9	H
Lamorna Gap	128.3	5000	-0.1	-23.4	H
Sea View	147	4830	0	-32.6	H
Bunkers Hill 4	256.6	815	0.5	-8.6	H
Carn Brea	282	2680	1.8	7.7	H
Boscawen Un Barrow 6	310	409	2.3	24.3	H
Boscawen Un Barrow 5	317.8	260	2.8	28.3	H

The target declinations have then been examined against the 14 key event declinations. These were the equinox, summer and winter solstice positions and the lunar major and minor positions.

Four of these seven horizon features give declinations which are close to key events. These are as follows

1. Boscawen-ûn Field Menhir has a declination of 28.9°+/- 1.04°coincides with the rising lunar major standstill of 28.15°.
2. The Lamorna Gap declination of -23.6°+/-1.04°this coincides with a winter solstice Sunrise of -23.9°.

3. Boscawen-ûn Barrow 6 of 24.3°+/-1.04°coincides with a summer solstice Sunset of 23.9°.

4. Boscawen-ûn Barrow 5 at 28.3°+/-1.04°coincides with a setting major lunar standstill of 28.9°.

All fall within the error range of +/-1.04

Two other alignments Bunkers Hill East and Chapel Carn Brea at -8.6° and 7.7° do not seem to have any obvious alignments. Although it is worth mentioning that the Sun would set in these positions almost equally either side of the equinox.

The final position at -32.6° does not match any key event. This is understandable as it is another natural feature, a small sea view gap just south of the main sea view of Lamorna cove.

Considering the results

Out of the four celestial alignments found from the targets, two were for solar events and two were for lunar events. The directions of the alignment are shown below.

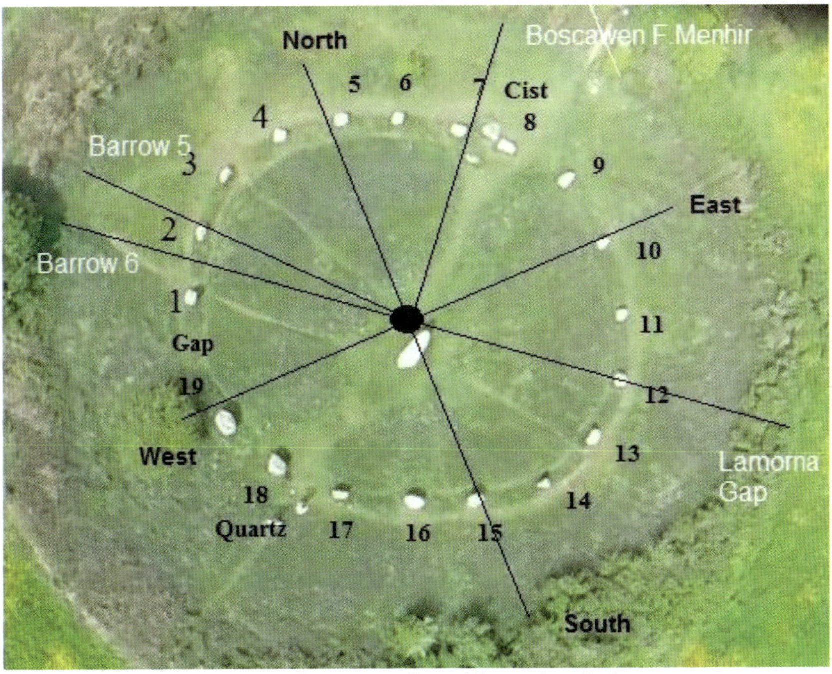

Alignments found from the circle

Solar Alignments

The winter solstice Sunrise occurring from the Lamorna gap is perhaps the one of the most credible astronomical alignments at the site.

Winter solstice Sunrise 2016

The Lamorna gap a natural feature that was important for sea access. From the position of the stone circle it would have been a subtle sea view even if the hedgerows were removed, due to the small change in altitude between the circle and the gap. An increase in altitude at the top of Creeg Tol shows just how much this view changes.

Lamorna gap from Creeg Tol (2016)

The Lamorna Gap alignment would have been at its most impressive when the sea levels were at their highest amplitude, from the timing of the semi-diurnal high tides, and the Moon at perigee spring tides. At the highest tides, the target would have covered near 10° of the horizon, making it a large area for an event to hit. Nevertheless, it makes a credible candidate for an astronomical alignment, as when sea levels were low this amount would have been reduced and closer to the measured 23.6° +/-1.04°. It should be considered as a primary alignment, as the circle would have been built here to align winter solstice with this natural feature, other features such as barrows and menhirs could have to be positioned later than the circle. If the circle was moved further up the hillside the Lamorna gap alignment would remain but the framing of Chapel Carn Brea would be lost, limiting the positioning of the circle to its current location. Boscawen-ûn with its northern hill and far-reaching southern horizon and a sea view is typical of many classic stone circle sites.

The second solar alignment is to a barrow 6 which is no longer there and was identified from historic records. It is worth considering that this barrow may have been positioned after the building of the circle to feature the summer solstice Sunset/winter solstice line. There are a large number of barrows around the circle and the positioning of this may have been coincidental. Without further archaeological evidence this alignment is unlikely to be established beyond doubt.

Lunar Alignments

The other two identified events with targets are at major lunar standstill positions. Current research into alignments tends to consider only recumbent stone circles found in Scotland and South West Ireland having lunar alignments. Although a stone row called Merrivale at Dartmoor, Devon includes an alignment with the lunar standstill (Wood 1978, Kelley 2011). Importantly the lunar alignments are also positioned over 5° in azimuth away from the closest event (the solstices) so there can be no overlap in the positioning of the events. The lunar alignment with the Boscawen-ûn field menhir is the most credible. It is at 28.9°+/-1.04° which coincides with the rising lunar major standstill of 28.15°.

Just the tip of this menhir is visible from the site. The menhir would be positioned slightly to the north of this position, although this menhir was once a possible pair which could have looked as below. Theoretically, the two menhirs could have framed a rising Moon but only further archaeological analysis could confirm the position of this pairing.

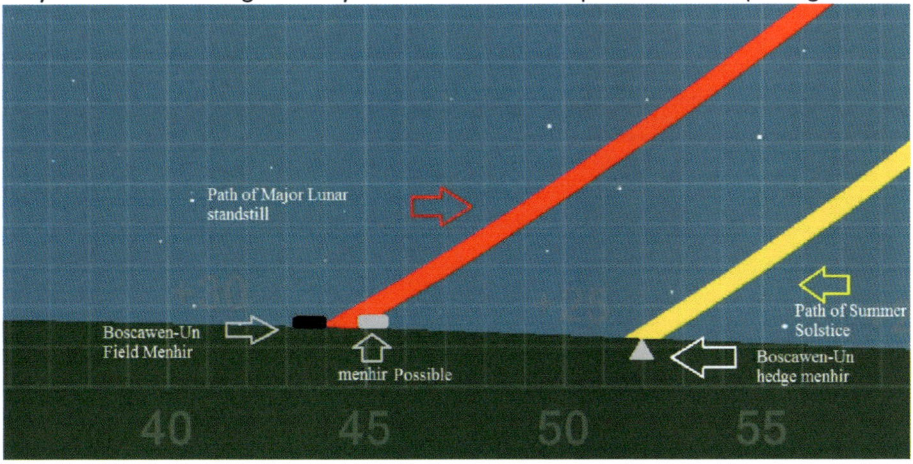

Boscawen-ûn Field Menhir lunar alignment Credit HORIZON

The second lunar alignment was also a major lunar standstill with barrow 5 on Creeg Tol. It is a possibility that a barrow was placed in the lunar standstill position or that the circle was built knowing that the lunar standstill position was at this granite outcrop.
There are a large number of solar alignments within Cornwall (Barnatt 1982), but how credible are lunar standstill positions at Boscawen-ûn? Evidence from local myths and legends give Boscawen-ûn a lunar link. Edmonds early 19 stone Metonic link could mirror the 18.61-year lunar standstill cycle. The circles' quartz stone is linked in local mythology to the Moon (Cooke 1986). Trade links between the local community have been shown to reach as far as Scotland and the transfer of knowledge about the lunar standstills could well have occurred. Any lunar alignment without precedence in Cornwall is difficult to decisively claim. But finding two such alignments at one site both for major lunar standstill positions adds value to the idea.

Other site considerations

Sunrise summer solstice

The hedgerow which surrounds the site makes it impossible to image the horizon at summer solstice Sunrise. The Sun was imaged 2 days before the summer solstice. The image shows the Sun rising in the NE, this image was taken at 4.31UT. The weather was cloudy.

Summer solstice Sunrise 2016

Lockyer suggested two alignments for the summer solstice. The Blind Fiddler was discounted due to its distance. But there is a more localised stone – The Boscawen-ûn hedge stone. It has an azimuth of 52°.5. This stone is not visible from the site but is relatively local at 665 meters. Visually when stood behind the quartz stone the Sunrise is behind the burial cist as suggested by Cooke (1986).
Cooke suggested the Sunrise on the summer solstice is in line with the cist. Even standing behind the quartz stone, the cist is too localised to be used as an alignment tool. But including Lockyer suggestion of an extended further alignment with the Boscawen-ûn hedge menhir. At azimuth 52.5° there is an alignment which forms between the cist and quartz stone.

The horizon has an altitude of 1.2° using

$$Sin\delta = sin\lambda sinh + cos\lambda coshcosA$$

This calculates a declination of 24.9°

This is the exact position of summer solstice Sunrise in 2500BCE.

Lockyer was correct in his calculations, but when considering this alignment, it must be noted that the stone is well out of visual range. It would be difficult to draw a conclusion currently on whether the Boscawen-ûn hedge menhir was positioned to indicate summer solstice. This line can be seen on the following image, which also highlights the distance involved. If it was used in this manner there would most likely be another marker between the circle and the hedge stone, or the observer could position themselves midpoint between the two. Although there is no conclusive evidence that this was the case, there is a possible fallen menhir nearby in a hedgerow between the features which could have linked these features.

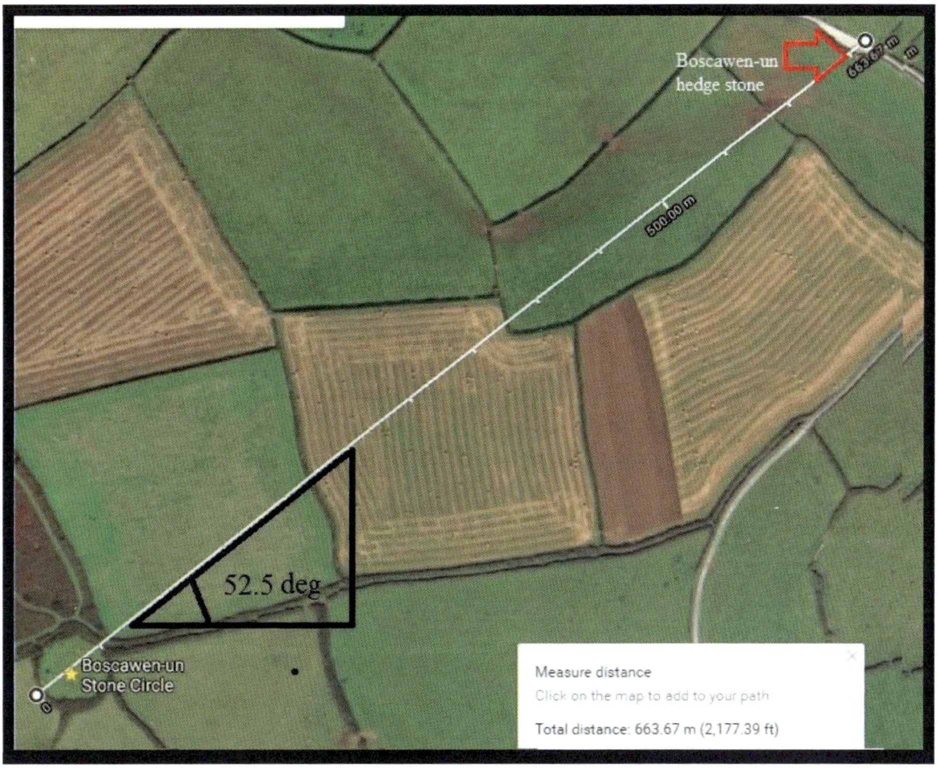

Boscawen-ûn Hedge Menhir alignment - Credit Google maps annotated Kennett

Another solstice feature – The central stone carving

A carving on the NE side of the leaning stone has a possible astronomical alignment (Cooke 1987). The manner in which the stone leans means this aspect is on the underside of the stone and for most of the year would be found in shadow or darkness. One time when this aspect is reportedly illuminated is when the Sun reaches summer solstice.

Daily observations would be impractical due to the large hedge which obscures the local horizon. This obscures the actual rising position of the Sun in this direction. It is possible to

calculate how many days the art would be illuminated using an azimuth reading, and the horizontal altitude. The azimuth reading needs to be between the angle of north as measured from the central.
Measured on-site the art would be fully illuminated with an azimuth lower than 60°+/1% with the horizon at altitude 0.9°. Once the Sun moves beyond 60° the most westerly foot starts to fall into shadow. With all the rock art falling into shadow by the time the Sun rises with an azimuth greater than 85°.
At the summer Solstice Sunrise in 2500BCE the art will be illuminated for 35 minutes 26 seconds.
In 2500BCE the Sun fully illuminates for the last time on the Solstice+30 days.

It is also useful to note that the art can also be seen partially illuminated near the Sunset during the summer months. It could never be fully illuminated during the evening, as the second foot is positioned too far to the NE.

Partial illumination of the central stone art before Sunset (2016).

Cross quarter Days

There is far less evidence for cross quarter alignments as suggested by Lockyer (1906). Let's look at the cross quarter day alignment with the Trelew line first.

May Day and the Trelew line of Menhirs

The Trelew line is made of a number of stones which radiate out from the circle towards the south east of the site. The first stone in this line is visible from the circle. This line was discounted from the sample above as none of the stones make a profile against the horizon. The line has an azimuth of 110.7° and an altitude of 0.3° using
$\sin\delta = \sin\lambda \sin h + \cos\lambda \cos h \cos A$

Where λ = latitude, h = altitude and A = azimuth
(All figures are given to an accuracy of =/-0.5°)
a declination is calculated at -12.87°.
Lockyer measured the Trelew line with an azimuth of 66°S and altitude of 1°. And a declination of 14°53S or -14.88°, this is a difference of 2° between our calculations and that of Lockyers.

Trelew standing stone.

In 2500BCE the November Sun would rise with a declination of -12.87 so it is not a cross quarter alignment. Lockyer's November Trelew line alignment is therefore incorrect and the line does not align to a November Sunrise. If you were to calculate the Sunrise in 2500BC it would infact align itself to a mid-November Sunrise around the 15th of the month.

Samhain Sunset

Samhain Sunset occurred in 2016 on 2nd November 16.48.50UT
Burl and Barnatt both suggested an alignment with Quartz stone. The image highlights how it wouldn't be an accurate way to pinpoint the date. The quartz stone is too close to the observer and any movement would change the setting position.

Samhain Sunset 2016

It is unlikely that it was used as an accurate calculator for a cross quarter day. It could be considered if another stone was used as a marker, but there is no evidence of this currently. Stone 6 on the opposite side of the circle has interesting notches in the top of it and this event was imaged through these notches but it also would have been out of line in 2500BCE as from these notches the Sun would move further behind the central stone.

Samhain Sunset through notch on stone 6. 2016

The suggestion by Barnatt and Burl that the quartz stone was positioned to mark Samhain is interesting, but it would not have been an accurate way to pinpoint a date, just a general positioning, it could mark the approach of winter when the Sun moves past the quartz stone. It does not mark Samhain as the position of the observer would radically change any observations made. The value of the onsite photographs played a real part in disproving this alignment.

Lughnasa and a rolling Sunset?

The Sunset was imaged on 8th August 19.34.30UT. The seeing was excellent and weather cloud free. In 2500BCE the Sunset would still be north of Chapel Carn Brea, it would not create a rolling Sunset down the side of this hill.

Lughnasa Sunset (2016)

The eminent archaeoastronomer Clive Ruggles claims that the division of the solar year into cross quarters is difficult to prove and has been disproved on numerous occasions. He considers that numerous dates in the cycle could hold significance in the calendars in the past. We also must consider that people were not limited to a solar cycle and that many cultures use different methods of deciding what is a year and how to divide it into sections.

Central stone position

The stone isn't positioned in the centre of the circle, but slightly off centre. It currently leans with an angle of 45° although this angle may have changed. With a bird's eye view the stone is positioned on an azimuth angle of 58° to 238°. The red line which runs through the central stone towards the NE leads to a large smooth recumbent stone. It is unusual due to its location near the circle. There are no other large stones within close proximity to the circle. To draw conclusions about these outlying stone archaeologists would have to examine it.

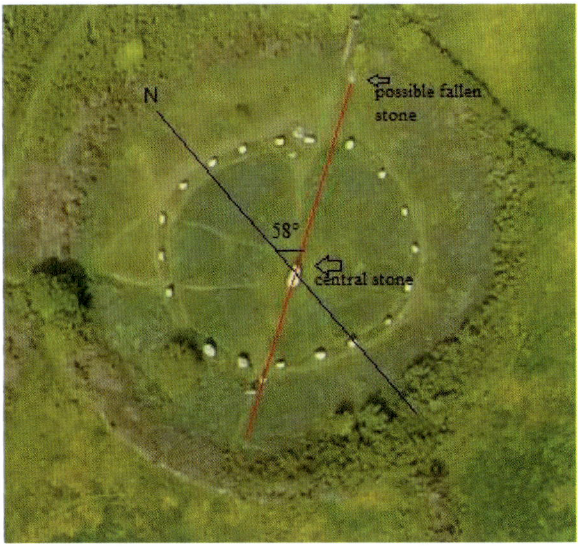

Central leaning stone, possible location of outlying stone.

The image below shows one problem when considering the leaning central stone. Theoretically what it points at depends on the observer. How they position themselves, how tall they are, and the distance they would choose to observe at would all play a part.

Viewing central stone alignments

We can calculate the declination of an object that proceeds past the highest angle of the central stone using the following calculation

To calculate the declination, we can use

$Sin\delta = \ sin\lambda sinh \ + cos\lambda coshcosA$

Where λ = latitude, h = altitude and A = azimuth

All figures are given to an accuracy of +/-0.5°

With an azimuth of 58° and an altitude of 45°

Sinδ = 0.810
δ = 52.8° or 52°48′
Arcturus is the closest +1 magnitude star to pass the top of the stone.
In 3820 BC Arcturus had a declination of +52°48′
In 3000BC Arcturus has a lower declination of +48°34′
If this central stone was aligned to Arcturus it means that the stone was positioned there in the mid-Neolithic period, but we must consider that there is no archaeological evidence that this is the case. It must be considered that the angle of the stone could have possibly reduced over time. This would change the results by moving the declination to an even higher value in the sky it moves further from Arcturus towards the NCP.
Could it point at the rising Sun at Solstice?
Rising with its lowest azimuth, the observer would need to be looking horizontally from a standing position and not looking at the pinnacle to see this. In -2500BCE the Sun would rise an angle of 50° (Ruggles 1999) it would reach an altitude of +4°47′47″ meaning that even at this stage the Sun would be too low and pass below the tip, so the central stone could not be angled towards the rising solstice Sun.

Lockyer and the Pleiades

Lockyer made two suggestions with the Pleiades. Lockyer's first alignment suggests it in a rising position with a distant cross and is not credible. The second suggests an alignment with a stone at 83°N or 277° As discussed within the historic suggestions this stone is not there. It could possibly reference the Bunkers Hill east at 256.6° or Chapel Carn Brea at 282°.
It is worth showing the issue with stellar alignments using this as an example
The image below shows an alignment between Chapel Carn Brea and the Pleiades. The alignment for the Pleiades setting here is correct for the epoch of 1500BCE, but earlier at 2500BCE it isn't, as the Pleiades would set further south to the feature. Care also needs to be taken when linking a date to match an alignment. Dating Boscawen-ûn from an alignment would be at best conjecture without the adequate archaeological evidence to confirm it.

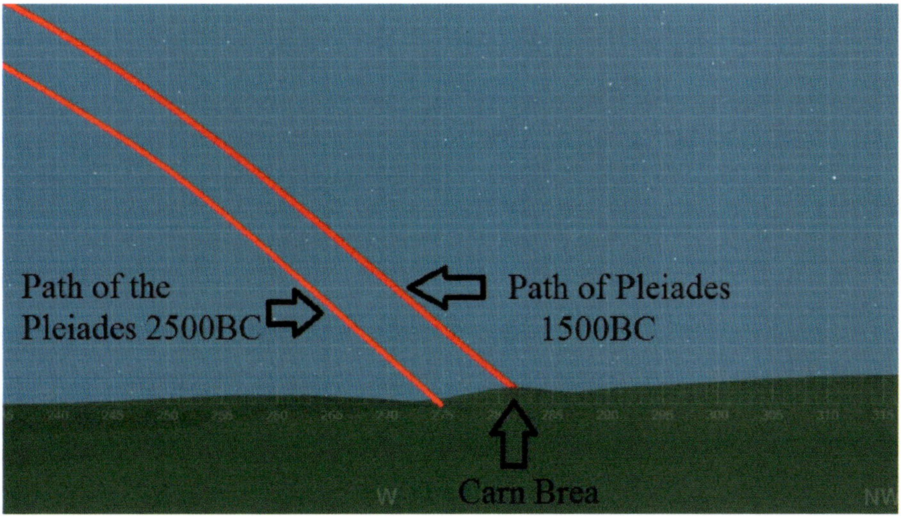

Changing setting position of The Pleiades Credit HORIZON

Furthermore, the Pleiades has a magnitude of 1.5 and Chapel Carn Brea has a height of 1.4°. Is it even possible to see it set over Chapel Carn Brea? Eta Tauri has a magnitude of 2.96 would need to be 3° above the horizon. So, taking into account the extinction of starlight it would not be fully visible even in the best of seeing conditions. Chapel Carn Brea could work as a setting position of the Pleiades but only between the declinations of 7-8.1° between 1800-1500BCE only with exceptional seeing conditions. This does not stop archaeoastronomers suggesting alignments with the Pleiades, particularly as its historically represented in art from Paleolithic paintings to the Nebra Sky Disc.

Conclusions

The positioning of Boscawen-ûn within the natural landscape has been done with a deliberate eye for astronomical events. The alignments discovered are positioned in the main to the northern side of the site and this may be due to the selection of objects on the horizon rather than below the horizon. More accurate dating of the site would lead to more conclusive results and also could lead to the suggestion of stellar alignments.

Solar alignments

There is an alignment between the circle and Lamorna Cove and the winter solstice Sunrise. The Cove would be a particularly important place for the local community as it allowed access to the sea. Peter Herring in 2014 suggested the Trelew line of menhirs running from Lamorna cove to Boscawen-ûn could have been direction markers to the circle from the cove This seems like a real possibility as the stones do not seem to have any astronomical alignment. The positioning of Boscawen-ûn Barrow 6 on the NW hill above the site at 24.3°+/-1.04° coincides with a setting summer solstice of 23.9°. Standing at the barrow looking SE it would also have faced the direction on the winter solstice Sunrise. The barrow no longer remains to confirm this.

The rock art on the central stone is illuminated in full and for the longest amount of time on the summer solstice. This seems to have been positioned deliberately as it is low down on the back side of the central stone. Although there is some partial illumination in the evening and later in the year the art on the stone would only have full illumination +/-30 days either side of the summer solstice. On the summer solstice, it would be illuminated for the longest period of time, removal of vegetation would see its duration 35minutes26seconds. It was impossible to say if the leaning stone was positioned to point at anything. The quartz stone, cist and pointed Boscawen-ûn hedge menhir form a line at the summer solstice Sunrise, winter solstice Sunset position. Care needs to be taken with this alignment due to the menhir not being visible (being just over the brow of the hill). There is no conclusive evidence for equinox or cross quarter days being aligned. Quickly moving between the circle and Creeg Tol, you can observe a phenomenon - a double Sunset or Sunrise.

Lunar major and minor

Creeg Tol would have a lunar major Sunset occurring at it every 18.9 years, the natural outcrop and possible barrow is at 28.3°+/-1.04° with a setting major lunar standstill of

28.9°. There is also a lunar standstill point with the Boscawen-ûn field Menhir. Results of 28.9°+/- 1.04° coincides with the rising lunar major standstill of 28.15°. Lunar standstill positions have had no previous precedence locally, the positioning of this menhir show an understanding of the lunar standstill positions by the builders of the circle. Without precedence and until further work is completed at other circles these alignments must be taken under advisement.

Stellar alignments are difficult to ascertain without more accurate dating of the site. But the positioning of the circle with the local high ground to the north would highlight the movement in circles of the stars above Creeg Tol such as Arcturus. Other menhirs, barrows and local features do not seem to have any astronomical significance.

The value of the historic observations was in the generation of ideas. Lockyer's work was the most scientific and proved remarkably accurate, even though the suggestions proved unlikely. Other historic suggestions held value, such as those made by Ian McNeil Cooke about the illumination of the central stone.

Boscawen-ûn Photographed through the year

The Sun changes its location on the horizon during a year. At the summer solstice, it will rise and set in its most northerly position. At the winter solstice, it will rise and set at its most southerly position. At the equinox, it rises virtually in the east and set virtually in the west. There has also been a small change of 1° of how far north and south the Sun rises and sets at the solstices. In 2500BC it would rise 1° further north for the summer solstice and 1° further south for the winter solstice. This will not make too much difference to the images as the Sun equates to about ½ degree in size, so it will be 2 solar widths at its maximum.

The Merry Maidens or Daens Maen.

The Merry Maidens is the easiest of the four circles to access. Located just of the B3315 west of Lamorna, there is a small parking area adjacent to the circle of the side of the road. It is a curious site with both similarities and differences to the other circle locations. The circle is located in a field surrounded by lush farmland, being on west Penwith's lowland southern plateau. It does not have a local commanding northern hill like many of the Cornish circle sites, instead it is located on the side of a southern ridgeway. Out of our four remaining circles it is located closest to the sea, but you would not know this as there is no way of viewing it from the site. The sea would have been to the south of the circle with the important Lamorna gap sea entrance just a short walk to the east of the site, but both of these are obscured from view due to the circle being located behind a ridge. The circle does have commanding and far reaching views to north-west and north. The distant hills to the north run from the most western point of the peninsula near Landsend and continue east towards Penzance. These hills Start with Chapel Carn Brea, through to Caer Bran and beyond. The circle is visible from both Boscawan-un and a very distant Boskednan circle (which is directly north of the site).
The circle itself is in a good state of repair. With nineteen stones forming an almost perfect circle and a small gap for a possible entrance. There are also a number of recumbent stones located in the field by the circle. How these were part of the circle complex is difficult to determine. But they are too close to the circle to be used as astronomical sightlines.

Adding to the complexity of the site there was another now removed circle in the field to the west of the site and a possible third to the south-east. The site itself is surrounded by a number of barrows and menhirs. The most famous and once most accessible is the

Tregiffian barrow. Sited next to the road a short walk west of the circle this barrow is well worth a visit.
The site was first examined historically by William Borlase in the mid-18th century and reported the second circle to the southwest of the site. This is known as the Boleigh circle and other than some large stones in the field where it was once located there are no remains of this circle. There have been a number of sketches and plans made of the site during the 19th century. In 1804 W and G Cooke drew an engraving of the site looking southwest towards St Buryan and the ridge line which extends from Chapel Carn Brea towards the north. The stones are depicted as being particularly large in scale compared to the sheep in the foreground!

W and G Cooke 1804. The Merry Maidens

In 1827 William Cotton included this site in his comprehensive book on the antiquities of West Cornwall. This is the first plan of the site, Borlase neglecting to draw this circle within his book Antiquities in 1754. Cotton's drawing and plan below give the view to the west of the site out towards St Buryan. What is interesting is that is shows a number of stones in the foreground which seem to be missing. With 16 stones standing in total.

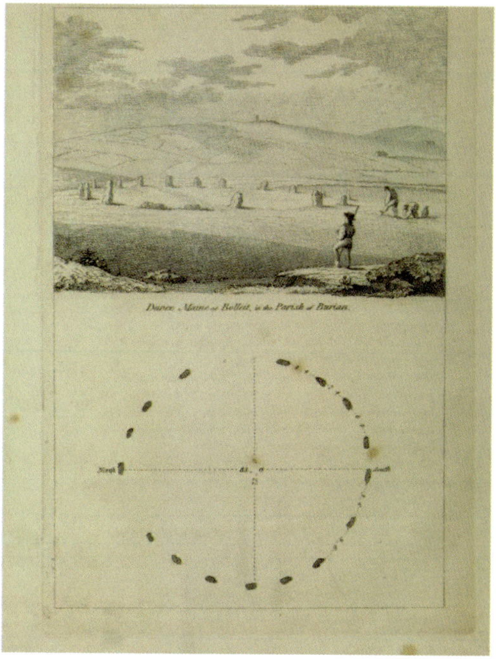

Looking SW St Buryan ridge
William Cotton (1827) Illustrations of stone circles, cromlechs, and other remains of the aboriginal Britons, in the West of Cornwall..

Richard Edmunds visits and sketches the site 30 years later. He also draws the site looking towards the West and northwest. He depicts at least 17 stones standing in the circle, he also shows one of the outlying stones to the left of the sketch. This is the first time that one of the outliers has been represented in an image.

Richard Edmunds 1857 Archaeologia Cambrensis
The Celtic and Other Antiquities of the Land's End District of Cornwall
Artist W Willes.

William Copeland Borlase gave us an insight into the local region around the site, drawing a plan of the circle and the local objects in the region. Many of these objects are still easily visible today. Such as the Pipers and barrows A and C. This image is published in 1872 in the text Naenia Cornubiae.

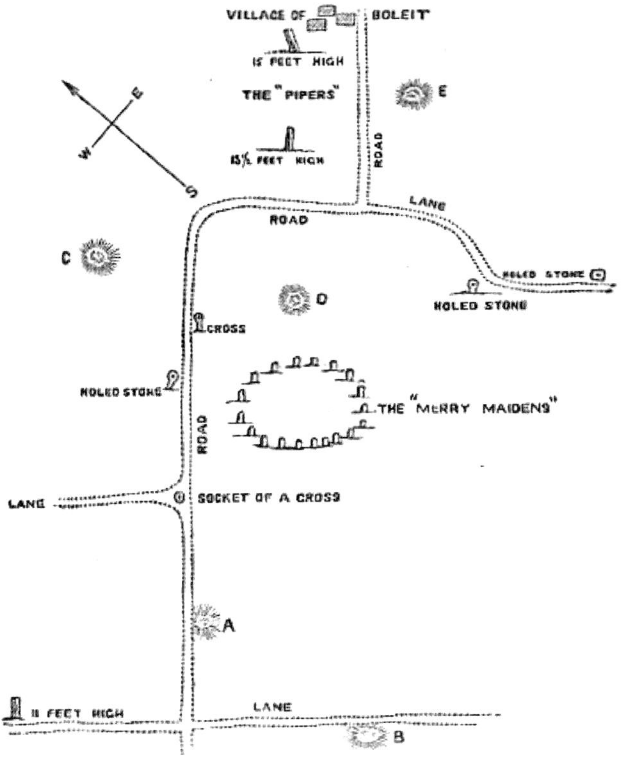

PLAN OF MEGALITHIC REMAINS AT ROSEMODDRESS AND BOLEIT.

William Copeland Borlase 1872

The Merry Maidens has proved a popular site with tourist and locals alike. An early photograph shows the site in 1905. It shows a site which is very reminiscent of what would greet a visitor today. A well-kept stone circle in a lush grassland field all enclosed by a hedgerow.

Looking NE towards the Pipers
Postcard of the Merry Maidens Stone Circle, St Buryan, c.1905 (The Merry Maidens, Pictorial Stationery Co Ltd, London, Autochrom, Peacock Brand, No. 1746)

Myths and Legends

The popularity of the site has added to the number of tales which are told about the stones. This has lead to a number of myths and legends being passed through the generations - the first extract is taken from Rambles in Western Cornwall by the footsteps of the giants by Halliwell-Phillipps, Publication date 1861.

"This is a very perfect: circle of nineteen stones which average about three feet and a half in height above the ground, the circle itself being nearly seventy feet in diameter. There are various country traditions which account for the existence of these stones. Some say that they were Maidens who were transformed into stones for dancing on the Lord's Day. Others assert that a man is buried under each stone. All, however, agree that the stones are placed there by supernatural agency, and that it is impossible to remove them. An old man at Boleigh, who informed us that a farmer, having removed two or three of the stones on one occasion, was astonished to see them in their old places the next morning, was evidently displeased at the account being inconsiderately received with a smile of incredulity. Another story respecting them is, that an attempt to drag them out of their places, although a vast horse or oxen power was engaged, utterly sailed, and that the cattle employed in the task fell
down, and shortly after died."

Western Morning news 1876 21st August tells a story of ancient people being turned to stone and a link to a Saxon King.

"The highly interesting group of ancient remains at and around Boleit were carefully examined under guidance Mr. Borlase. They consist stone circle of the Merry Maidens Rosemaddress; and the huge menhirion or monoliths known as the "Pipers" hard by; with, at some little distance, the curious Fogou, artificial cave at Trewoofe, built up by Britons

underground with blocks of stone. story with regard the circle and menhirion the former were young women who were turned into stone for dancing on Sunday, and that latter were men who piped to them, and who shared their fate. There doubt that the menhirion have a double character, that they are sometimes sepulchral, and sometimes memorials of events. Now, it is curious that while Boleit has been interpreted "house of slaughter," Mr. Borlase has found confirmatory of Dr. Ferguson's view of the purpose the circles, that there are still extant traditions battle at Boleit, where, indeed. It is said that Athelstan won his final victory over the Britons. In fact this legend has been expanded to say that the Pipers represent Aethelstan and Howel two of the leaders that fell in battle that day."

Another longstanding myth gives a link to King Arthur and was told here within the Western Morning News 21st Sept 1928

"From Paul the party will drive towards the Logan Rock, halting at Boleigh to view the very remarkable group of prehistoric monuments surrounding the stone circle called the Merry Maidens. No area in Britain is more thickly strewn with memorials of remote antiquity than this. The circle itself—only less fine than that of Boscawan-Un (the scene of the Gorsedd)—stands in a field. Near it are the two famous " Menhirs " or Long-stones called the Pipers and other stones of lesser note. At Boleigh there is local tradition of a great battle, and the name Rosemoddress, the " Heath of Modret,' has suggested that this was the fatal field where King Arthur was slain his treacherous nephew Sir Modret."

Historic alignments

Norman Locker originally visited the site at Easter 1905 and was to return in January 1906. These visits lead to the first of his papers on Cornish circles to be published in Nature magazine. Accompanied by local dignitaries he made measurements of the azimuth between the circles and the features he would consider for alignments.

Nature Feb 1906 Vol 73

He would consider 8 alignments to local stones and monuments. None of these were natural landscape features. These are shown on the map below

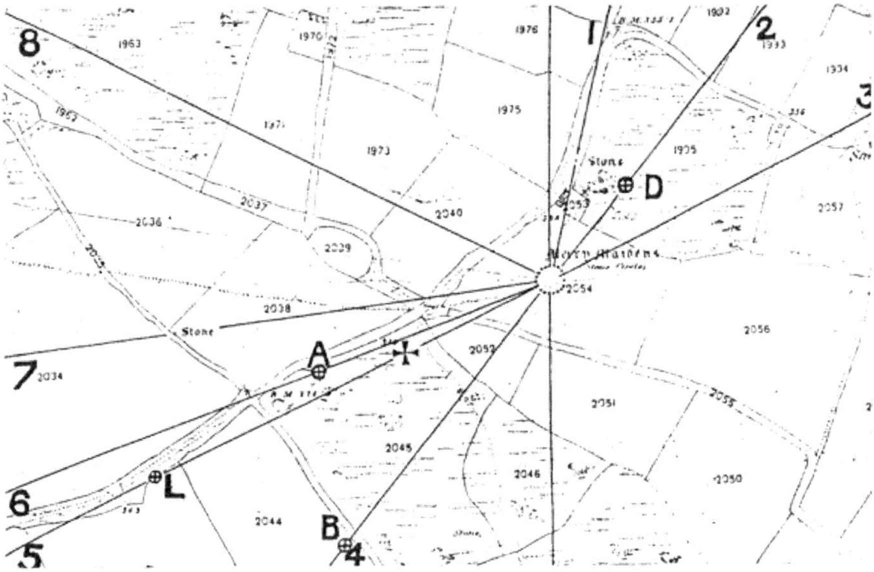

A is Tregiffian barrow
B is another local barrow
D is one of the pipers
L is another local barrow

Lockyer's alignments are all outlined in the table below.

TABLE OF ALIGNMENTS.

Alignment.	Azimuth.	Hill.	Decl.	Sun or Star.	Date.	Mark.
1	N. 11° 45 ′E.	20′	38° 46 ′N.	Arcturus (warning August)	B.C. 1650	Stone in road.
2	N. 38° 25 ′E.	20 ′*	29° 58 ′N.	Capella (warning February)	2160	The Pipers and barrow.
3	N. 64° E.	½°	16° 21 N.	May year . . .	—	Fougou.
4	S. 38° 22 ′N.	20′	30° 27 ′S.	Pipers line . .	—	Barrow B.
5	S. 64° W.	20′	16° 26 ′S.	May year (February-November setting)	—	Barrow L.
6	S. 69° W.	32 ′*	13° 18 ′S.	Antares (warning May)	1310	Stone cross on hill and Barrow A.
7	S. 81° 35 ′W.	32 ′*	5° 24′ N.	*Reversed line.* Pleiades elev. ½° (warning May)	1960	Goon-Rith.
8	N. 64° W.	42′	16° N.	May year (May eve setting)	—	St. Burian Church.

More modern examination of the site happened through John Barnett and John Mitchell, who agreed with many Lockyer's suggested alignments. These were published along with more in a map which appears in Ian McNeil Cooke's "Journey to the stones". This map also includes two additional alignments suggested by the team at Meyn Mamvro. A total of 11 alignments are suggested.

Site	Azimuth	Distance	notes
Nun Carag Cross	11.09	90	Lockyer alignment 1
Pipers I	38.02	327	Meyn Mamvro I
Pipers II	38.69	422	Lockyer 2 and 4 Michel 1
Boleigh - bronze age barrow	64.53	621	Lockyer 3 and 5
Tregurnow stone circle	82.09	471	Barnatt 1 (Lockyer 7 extended)
Boscawen-Ros East	216.1	721	Meyn Mamvro I
Boskenna Tumuli	219.2	399	Lockyer 2 and 4 (barrow B)
Boscawen-Ros West	219.63	754	Lockyer 2 and 4
Tregiffian cist	232.95	263	Lockyer 3 and 5 (barrow L)
Merry Maidens /Boleigh circle	234.16	83	Lockyer 3 and 5
Tregiffian Barrow (cup marked)	250.62	240	Lockyer 6 (Michell 2)
Gun Rith	264.15	323	Lockyer 7 (Barnatt 1)
St Buryan Church	2560	202	Lockyer 8

It is worth considering that two of Lockyer's alignments are to monuments which date to much later than the circle. These are Alignment 3 and Alignment 8.

Alignment 3 was originally to the Fogou at Boleigh but has been superseded by the suggestion it is to a Bronze age barrow at the same location.

Lockyer 8 refers to Buryan Church which he suggested could have stood on a much earlier sacred site, but without archaeological evidence to back this up this cannot be suggested to be conclusive.

Four of Lockyer alignments are to stellar objects which without a clear date for the site would also be difficult to conclusively prove (see Boscawen-ûn Pleiades discussion in Chapter 2).

Three more alignments have been suggested for May Day. And no astronomical alignment has been suggested for the Pipers.

These alignments do not give astronomical explanations beyond those suggested by Lockyer. It seems that the circle of the Merry Maidens has tried to hold onto its secrets more than most.

Current Measurements

Recent onsite measurements have been taken and these have a list of local sites and natural site lines as follows

Site	visibility	Azimuth	Distance	notes
Boleigh	N	38.23	519	
Boleigh Holed stone	N	41.14	549	
Boleigh Tumulus	N	61.12	677	
Lamorna	N	56.51	1.5	
Boleigh - bronze age barrow	N	64.53	621	Lockyer 3 and 5
Tregurnow stone circle	N	82.09	471	Barnatt 1 (Lockyer 7 extended)
Boscawen-Ros East	N	216.1	721	Meyn Mamvro I
Boskenna Tumuli	N	219.2	399	Lockyer 2 and 4 (barrow B)
Boscawen-Ros West	N	219.63	754	Lockyer 2 and 4
Boscawen Ros Barrow	N	220.52	286	
Boskenna Round	N	231.38	804	
Lanyon Quoit	N	355.49	9150	
Nun Carag Cross	Y	11.09	90	Lockyer alignment 1
Pipers I	Y	38.02	327	Meyn Mamvro I
Pipers II	Y	38.69	422	Lockyer 2 and 4
Rosmodress barrow 2	Y	43.76	131	
Barnatts stone	Y	79.98	161	
Rosmodress Holed stone	Y	89.54	271	
Tregriffen Barrow (south)	Y	189.26	221	
Tregiffian cist	Y	232.95	263	Lockyer 3 and 5 (barrow L)
Merry Maidens /Boleigh circle	Y	234.16	83	Lockyer 3 and 5
Tregiffian cist 2	Y	243.13	128	
Tregiffian Barrow (cup marked)	Y	250.62	240	Lockyer 6 (Michell 2)
Gun Rith	Y	264.15	323	Lockyer 7 (Barnatt 1)
Chapel Carn Brea	Y	304.92	5970	
Tregadwith	Y	315.43	1.37	
Bartinney Castle	Y	319.07	6130	
Rosmodress Holed stone	Y	351.24	76	
Mên-an-Tol	Y	354.45	10460	
Carn Galver	Y	355.49	11400	
Rosmodress Barrow	Y	358.21	202	Lockyer barrow D
Boskednan	Y	359.1	10610	

Once again these have been reduced to just the objects which are visible from the site. The declinations of these objects have been calculated and are in the following table below

Merry Maidens is situated at a latitude of 50.07° and a longitude of -5.59°

Name	Azimuth	Distance	Altitude	Declination	Horizon
Pipers I	38.02	327	-0.1	30.29	H
Chapel Carn Brea	304.92	5970	0.9	22.22	H
Rosmodress Holed stone	89.54	271	0.3	0.53	H
Bartinney Castle	319.07	6130	1.0	28.85	H
Carn Galver	355.49	11400	0.7	40.49	H
Boskednan	359.1	10610	0.8	40.73	H
Nun Carag Cross	11.09	90	0.5	39.54	L
Pipers II	38.69	422	-0.2	29.89	L
Rosmodress barrow 2	43.76	131	0.1	27.71	L
Barnatts stone	79.98	161	0.2	6.57	L
Tregriffen Barrow (south)	189.26	221	-0.3	-42.49	L
Tregiffian cist	232.95	263	0.3	-22.5	L
Boleigh circle	234.16	83	0.3	-21.83	L
Tregiffian cist 2	243.13	128	0.4	-16.54	L
Tregiffian Barrow (cup marked)	250.62	240	0.7	-11.75	L
Gun Rith	264.15	323	0.4	-3.44	L
Tregadwith standing stone	315.43	1.37	1.1	28.16	L
Rosmodress Holed stone	351.24	76	0.8	40.17	L
Mên-an-Tol	354.45	10460	0.7	40.41	L
Rosmodress Barrow	358.21	202	0.8	40.71	L

There are six targets which fall on the horizon. The locations reflect the far-reaching views from the west to the north of the site. Much of the south is obscured due to the localised southern ridge next to the circle. Although without the hedge it would be possible to see the holed stone of Rosmodress on the horizon.

Two of the calculated alignments are close to the declination points of the lunar major rising and setting positions.

Pipers I at a declination of 30.29° which is close to the northern most lunar major rising position. This will be discussed further in the following section.

Bartinney castle is in the lunar major northern most setting position at a declination of 28.85° it is very close to the actual declination of 28.15°.

Two more targets on the horizon are at the position of cardinal north. These are the distant hill of Carn Galver and the Nine Maidens stone circle shown in the image below (taken from Tregriffen Barrow.

A final target is the Rosmodress holed stone which is positioned at the equinox position to the east of the site. Could this be a sightline for the equinox rising and setting Sun over the circle? It is a real possibility that this may be the case. Apart from this one link to the solar cycle there does not seem to be any horizon features with the solstice around the circle. You can see in the image below that the summer solstice Sun would have always set on the ridge between Chapel Carn Brea and Bartinney.

Imaging the site approaching summer solstice Sunset 2017

The image was taken on June 10th, 2017, 11 days prior the solstice. This means that the Sun would set another solar disc width towards the north (right of the image above). We also could consider that the Sun has moved due the obliquity of the ecliptic. This would account for another solar width to the north. It is obvious that the Sun does not align with the crest of any of the distant hills in the image at the solstice Sunset.

The Pipers

One of the questions asked about the Pipers, queries if they are they visible from the site?

With modern techniques, we can measure the altitude and heights of the stones.

Altitude measurements
At the Merry Maidens the altitude is 102m. But there is a slight ridge in the direction of the Pipers which would act as a visual barrier to anything lower in the landscape.
At Piper 1 the altitude is 98 m, the height of stone 4.7m tall which totals 102.7m – just the height needed for the top of it to be visible.
Piper 2 altitude at the stone is 91 m. The height of stone is 4.2m (but leaning) This means that the height of the top of the stone is 95.2m, this means it would not be visible from the site.

Why is important to consider this? The Pipers lie on an azimuth which is close to the declination in the position of the lunar major north rising position. This is remarkably reminiscent of Boscawen-ûn field menhir which also lies in the same position at the Boscawen-ûn stone circle. The top of this menhir is also only just visible from the circle and possibly was one of a pair. Could the same double menhir tradition have been used at both sites? If so the set up would have been different. For instance, at Boscawen-ûn the menhir if a pair would not have been separated by the distance that is found at the Merry Maidens. Also, the second piper is not inter-visible and the first would need to have a flat landscape as any vegetation would have made it impossible to see. Finally, the Pipers have been measured at a declination of 30.29° which is over 2° difference to the Northern lunar major rising position (or 4 Moon discs in distance). So, it is very doubtful that they were placed to measure this astronomical event.

Another consideration about the Merry Maidens is that over long distances it is in alignment with the Boscawen-ûn circle. The lunar major southern standstill position falls on the ridge above the circle itself.

Merry Maidens to Boscawen-ún has a -29.9° declination which places the Merry Maidens in the position of the Southernmost lunar major rising position of 29.85°. The adjacent circle would have had a declination of 30.1°

The following image shows how the circle sits in the lea of the ridge which forms the southern coastline, this image shows the Merry Maidens just to the right of the farmhouse on the ridge and the pipers standing towards the left of the image. The picture was taken from Boscawen-ûn stone circle.

Far reaching views from Boscawen-ûn showing the Pipers and Merry Maidens

The Moon rising at its most southerly position over the stone circles of the Merry Maidens and Boleigh circle would have made quite a spectacle when viewed from the Boscawen-ûn complex.

Adding to the whole idea of a lunar aligned landscape is the fact that Tregriffen barrow has a number of cup marks on it. The use of cup marks has been linked to the lunar cycle and further consideration may be given to the location of cup marked stones within the locality.

The Merry Maidens is located within a densely packed ceremonial landscape with many monuments which could have been included for astronomical consideration. Further work may wish to consider the role of holed stones within the local environment and the positioning of the other barrows.

Boskednan (Nine Maidens)

"Boskednan Circle, with its fallen pillars and its barrows overgrown with weeds, located in a dreary moorland that is thoroughly in harmony with it. Treeless, sown with innumerable hoary blocks of granite, swept by blasts from the open Atlantic, which here are always moaning and sighing, and frowned over by hills that are only little more savage than itself, the upland where the Circle is situated looked particularly desolate this morning and the ruined Circle, with its wild surroundings, impressed imagination as few other antiquities of the district have been able to do."

Excerpt taken from "In the land of granite and gorse - notes of a tour through west Cornwall in the summer of 1886". J. H. Pearce. Published Cornishman 16th July 1908.

The first time I visited this circle I got lost, horrendously lost. I had approached the moor from the small carpark at the chapel on the road from Madron, having stopped at Lanyon Quoit along the route from Penzance. The moor doesn't look that large, but it is deceptively bigger than it first looks. The day I visited was early spring and misty adding to my later extended hike. The first stop on this well-trodden route is Mên-an-Tol, Boskednan's better known neighbour. Mên-an-Tol is a fascinating site which will be discussed in the following chapter along with some of the other holed stone sites in Cornwall. Following this it is worth a visit to the inscribed stone Men Scafa in a field adjacent to the pathway. Once you have completed this stop it is a gentle climb to the top of the hill where Boskednan is located at SW 436351, Latitude 50.160664 Longitude -5.593850. Don't get this wrong, Boskednan is a relatively easy place to find particularly if you walk from the direction of Ding Dong mine and if you are heading in from Mên-an-Tol there are several paths which lead up to the top

of the hill. The location of the circle is unusual as all the other circles we have considered are in the lea of large hills rather than being on high ground themselves. It was possibly this which got me confused and on the misty day turned about in my search for the circle. Once you reach the circle you can see on a clear day impressive and far reaching views.
The first documentation of the circle exclaims it was in a ruinous state, over the years since then many stones have been re-erected. In the mid-18th century William Borlase sketched 13 erect stones and the possible original 19-22 stones. The sketch is shown below it has been taken from his book 'Antiquities, Historical and Monumental, of the County of Cornwall' published in 1769. He records it as being similar to the circles in the peninsula and thinks that originally there was 19 stones.

Borlase 1769 Antiquities, Historical and Monumental, of the County of Cornwall

By the time William Cotton visits in 1827 he finds a monument which he considers originally had 19 stones, of which 9 are still standing. There are another 7 lying recumbent and holes of other missing stones. He drew a plan which is shown below.
Interestingly he has not drawn the dominant Carn Galver, instead the image is looking out to the far-reaching views in the west.

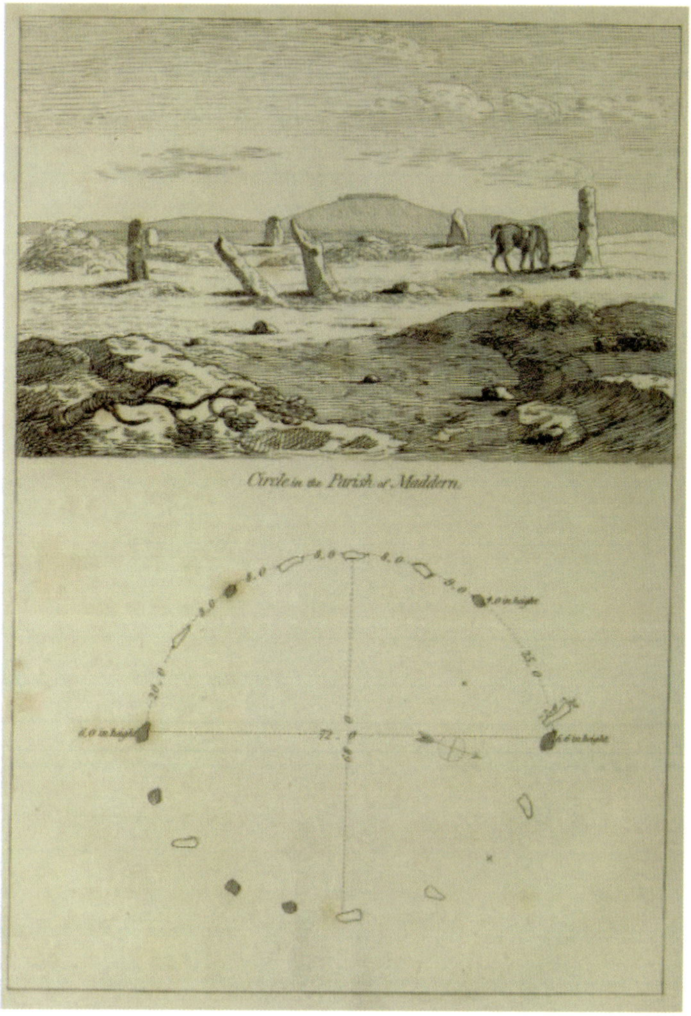

William Cotton 1827 Illustrations of stone circles, cromlehs, and other remains of the aboriginal Britons, in the West of Cornwall.

In 1868 H.P. Blight writes about Boskednan within the Gentleman's Magazine (March 1868) in an article called "*notes on Stone circles*".

He claims that Boskednan is not a true circle erect, but "the remains of a base-circle of a larger proportion of a "twin barrow". Although this is no longer considered the case he interestingly goes on to draw 19 stones of which 12 are standing. Blight was a very well-respected artist and therefore it is likely that he recorded a true representation of the circle at that time. When comparing the diagram that he sketched of the stones to that which Borlase drew 100 years prior they are remarkably similar in orientation and spacing of the stones.

Fig. 9.

His plan also shows 19 stones and the attached smaller barrow. He considers that the two central stones A and B are the remains of a cist.

HP Blight (1868) Notes on Stone circles.

On the 26th July 1872 the site was visited by William Copeland Borlase and he drew the circle in and this drawing was published in 'Nænia Cornubiæ'. William Copeland Borlase noted how many circles had barrows or Cairns which were near the circle. The following is an account from Borlase of what he discovered.
"This circle consisted in Borlase's time of nineteen stones thirteen of which were then erect with a detached long stone forty three paces to the northwest 1 Eleven of these stones are all that now remain seven of which are standing while four are prostrate Of those now standing the average height is four feet with the exception of a fine menhir 2 on the northern side which measures 6 feet 6 inches above the surface of the ground Immediately opposite this at a distance of seventy feet on the southern side of the ring stands a flat faced stone about four feet high the base of which is surrounded by the outskirts of a Cairn which thus is made to cut the circumference of the circle at this point So large a portion of the stones composing it have been removed that neither its height nor circumference can now be determined"
He continues his description saying how a trench was dug and a discovered cist opened which revealed a cremation and an urn.
He also mentions two further local Cairns
"some 200 yards to the north west of Boskednan Circle On the top of one of these stood in all probability a Kist Vaen mentioned in Borlase's MSS Paroch Mem p 9 .... Another large but dilapidated Cairn is to be found at distance of 200 yards east of the same circle 1"
In 2004 work was undertaken to repair the circle. Along with other works three long-fallen stones were re-erected to their original sockets

Currently there are only 11 stones and of the surviving stones, six are upright, two are leaning and three have completely fallen. The circle itself is about 21.8m in diameter, with the stones placed evenly around the circle. All the stones are granite and the tallest is 1.2m in height.

The most striking sight on arrival is the dominant granite outcrop of Carn Galver. This can be seen to the right of the image above. This image shows the northwest and sees a more localized ridge of White Downs with the Watch Croft outcrop visible from the circle leading back to the Carn Galver outcrop. The Sun is setting between Carn Galver and Watch Croft on the 2017 midsummer solstice.

The circle also has many further landscape features on the horizon. What is noticeable but subtle is a northern hill forming a ridge which extends towards Little Galver. On a clear day, there is a distant sea view to the NE. Following round to the south, the distant granite outcrop near Zennor which forms part of Treen Common, can be seen in the image below.

The view is lost towards the Southeast due to a rise in the local landscape. This can be seen in the picture below which is looking towards the south of the site. The man is stood on the local barrow which is adjacent to the circle.

Taken by Gibson and Sons in Penzance prior to 1906.

Once past this localised barrow and small hill the far-reaching views continue into the southwest, where the ridge with Rosmodress is visible and a far flung Merry Maidens may have been seen to the south. Past the modern village of St Buryan you can see Bartinney hill in the distance. These distant views extend round from the southwest through to the west where the Longstone Bowens Menhir is visible just south of Carn Kenidjack. This can be seen in the image below, with Bartinney labeled on the left and The Longstone and Carn Kenidjack to the right of the image

A number of natural features have been outlined in the 360 tour of the Boskednan horizon. These will be compiled and measured in the following section along with any objects positioned in the landscape could have been positioned to form orientations with the circle.
Local features and their locations.
Following are a list of sites which are visible from the circle. These are the sites that could form alignments with the circle. This list has been taken from a longer list of local sites, which has been reduced to just the monuments and landscape features which are visible from the site.

Site	Azimuth	Distance (m)	Altitude (m)	Declination	Horizon
Bosporthennis Cairn 3	1.93	528.27	1	40.81	L
Bosporthennis Cairn 1	4.29	556.21	0.8	40.5	L
Bosporthennis Cairn 2	8.93	517.33	0.6	39.86	L
Zennor Cairn	38.44	4450	0.5	30.55	H
Bronze age Cairn east 2	63.87	592.47	0.1	16.47	L
Bronze age Cairn east 1	64.52	231.09	0.1	16.08	L
Bodrifty cup marked stone	71.21	978.85	0	11.91	L
Mulfa Quoit	80.05	1750	0	6.41	H
Bronze age Cairn east 3	86.32	546.26	0	2.36	L
Treddinick Bronze Age barrow	92.61	754.82	-0.1	1.75	L
Treddinick stone circle	109.61	1.04	-0.5	-12.81	L
Bartinney Castle	211.65	7030	-0.1	-33.14	H
Boswens Menhir	233.16	5090	-0.2	-22.42	H
Carn Kenidjack	252.22	4570	-0.6	-11.75	H
Boskednan Cairn 11	262.64	261.7	-0.3	-4.94	L
Men Scarfa	278.91	751.58	0.6	6.16	L
Watch Croft menhir	287.48	1540	1	11.88	H
Boskednan Cairn 5	298.09	45.36	0.5	17.96	L
Rosemergy grave	300.06	857.27	0.7	19.43	L
Boskednan Cairn 4	310.23	159.9	0.2	24.61	L
Boskednan Cairn 3	311.62	181.04	0.2	25.35	L
Boskednan standing stone 7	312.18	235.78	0.2	25.73	H
Boskednan standing stone 6	312.39	48.21	0.2	25.79	L
Boskednan Cairn 1	312.26	235.8	0.3	25.78	H
Carn Galver	319.83	1009	0.5	29.75	H
Boskednan Cairn 2	321.27	209.89	0.2	30.16	L
Little Galver	347.83	925.5	0.5	39.25	H
Boskednan Cairn 8	140-170	adjacent	0.2-0.4	average - 33.42	H

There are many local barrows within the region and these have been numbered and are their general locations are shown on the images below. Some of the barrows are not visible from the site so have been removed from the table. A large cluster of barrows appear to the NW of the site and more occur to the south and west of the circle.

To the north of the site is a small hill on it are three Cairns called Bosprthennis - two form a profile on the horizon.

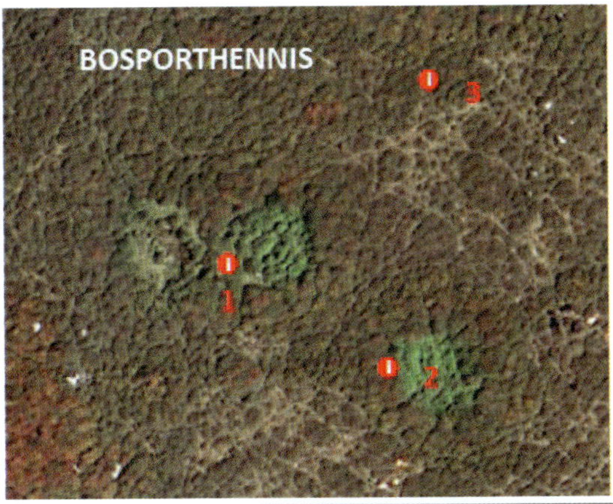

Bosporthennis Barrows to north of site
(Annotated Google maps)

To the NW of the site are a large number of barrows and two standing stones these are all called Boskednan and numbered in the image below.

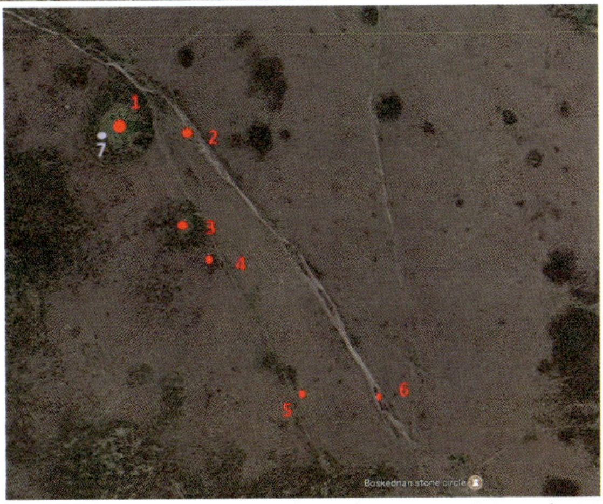

Boskednan barrows to NW of site (Annotated Google Maps)

South West of the circle are a number of barrows for the purposes of this project these have also been called Boskednan and numbered

Boskednan barrows SW of site
Locations have been placed on images taken from Google maps

The list of visible sites has been further reduced to those which are visible from the site and then to those which form a profile against the horizon. These are given in the table below.

Site	Azimuth	Distance (m)	Altitude (m)	Declination	Horizon
Zennor Cairn	38.44	4450	0.5	30.55	H
Mulfa Quoit	80.05	1750	0	6.41	H
Bartinney Castle	211.65	7030	-0.1	-33.14	H
Boswens Menhir	242.16	5090	-0.2	-22.42	H
Carn Kenidjack	252.22	4570	-0.6	-11.75	H
Watch Croft menhir	287.48	1540	1	11.88	H
Boskednan Cairn 1	312.26	235.8	0.3	26.28	H
Carn Galver	319.83	1009.58	0.5	29.75	H
Boskednan standing stone 7	312.18	235.78	0.2	25.73	H
Little Galver	347.83	925.5	0.5	39.25	H

Lunar major northern setting position is at 28.9°, Carn Galver is 29.15°
Lunar major northern rising position is at 28.9°, Zennor Cairn is 30.55°
Boskednan has been positioned with lunar alignments to natural features on the horizon. These are both lunar major northern most rising and setting points.

There are two possible solar alignments the first is with the local Barrow 1 towards the NW of the site. This has a declination of 25.78°, and the stone which is part of the barrow but stands prominent on the western side (standing stone 7) is at a declination of 25.73°. These are both close to the summer solstice setting position of 2500BC which would have happened at 25°.

The second is a solar alignment at the limit of the error margin, it is the position of the Boswens Menhir with the winter solstice sunset. It has been measured at -22.42° which is close to the winter solstice sunset of -23.9°, but just outside the error margin of 1.08°. The sun has been imaged in this general location near winter solstice below. Could this standing stone form a marker for a processional route undertaken on the winter solstice between Boskednan and the Tregeseal circle site? This will be returned to within the next section about The Tregeseal circle complex.

Carn Kenidjack and Watch Croft are also at an equal setting position either side the equinox. At -11.75° and 11.88°. This would place the Sun as a setting object at these positions at approximately 30 days before the spring equinox and 30 days after the autumn equinox.
The barrows 1,3,4 and 5 form a line and are positioned to form a line which leads all the way to Carn Galver. This line is on an azimuth of 326.17° which corresponds to a declination of 32.42°.

Carn Kenidjack is a distant but distinctive hill from Boskednan

If we are to consider stellar alignments but it is interesting to note that Little Galver and the Ring Cairns to the north of the site have been positioned in a location which would consider a sweeping bottom place of northern circumpolar stars.

Cardinal positions

As per many of the stone circle sites there is a northern hill. This hill isn't as dominant a feature as at many of the other circle sites in Cornwall, but is worth a mention nevertheless. It has a number of ring Cairns on it which are called the Bosporthennis Cairns within this report.
Interestingly the Merry Maidens stone circle is positioned directly south of the site. It isn't visible due to the large localised barrow attached to the circle. But standing on the barrow you could see this complex region over 12km away directly to the south.

Tregeseal stone circle complex

To the north of Boscawen-ún is the Tregeseal circle complex. This is called a complex as originally there were a maximum of three circles at this site, although today only the eastern circle remains. It is sometimes known as the 'Dancing Stones' and has even been known as the 'Nine Maidens'. The eastern circle is situated in a similar manner to Boscawen-ún in the landscape with both circles being found in a liminal location at the base of a local northerly hill. This time the northerly hill is called Carn Kenidjack. There is also a sea gap, but this times to the Southwest rather than the Southeast at Boscawen-ûn. The landscape is much more open. It is no less secluded but dramatic open moorland stretches out to the North of the site and the lowland farm land is capped pretty immediately by localized ridges to the south. The sea gap is far more pronounced being in a broad deep dip in the landscape. Here framed on a clear day is the Isles of Scilly, 26 miles in distance. Most people approach the circle from Stoney lane where there is nearby parking down a long winding lane. The approach from the north is much more dramatic. Parking is off the North Road near the Trewellard junction you can access the moorland site area via a lane which runs by Boslow Farm. This takes you to the pronounced and locally recognisable Carn Kenidjack. If you do walk this way there is a lovely example of a barrow called Portheras common barrow as you start down the lane to the farm which is well worth a stop. It's also worth noting that this barrow has been aligned with summer solstice Sunrise and Chun Quoit.

The view from Carn Kenidjack over Tregeseal

Carn Kenidjack is one of the most conspicuous natural features within the West Penwith peninsula. A drawing of the Carn by J.T. Blight below comes with the following description.

"the rocks on the summit assume the most fantastic forms; and an arrangement irregular and strange.... Yet there is an attraction about this hoary Cairn which must render it of interest to all. Its solitary and desolate aspect is one feature; a deep silence and mysteriousness hang over it." The drawing below shows the Carn looking towards the southeast the same way you will approach it if coming on foot via Boslow Farm. It's difficult to tell if its depicting the full Moon or the rising winter Sun.

CAIRN KENIDZHEK.

Image by J. T Blight A week at Landsend 1861

Taking a moment to stop at Carn Kenidjack is always worthwhile the views are far reaching. Looking directly south from Carn Kenidjack you can see Chapel Carn Brea. Although 4.25km in distance away you can still spot the Scillonian barrow on it's top. Chapel Carn Brea is framed by the more local ridge of the A3071 and Bartinney Castle. There is a sea view is to the west, with the town of St Just nestled in the bottom in front of it. If the weather is consenting with a clear day, you can see the Isles of Scilly rising out of the sea 26 miles away, its long golden beach providing a mirage like a giant distant wave.

In front of these sea views you can see the remaining stone circle; Tregeseal East. The second circle, Tregeseal Central, has now succumb to a neighboring farmers field and a possible third, Tregeseal West, having left no trace other than as a crop marks on an old aerial photograph. The area is heath-land, covered in gorse, bracken and heather. It's a haven for wildlife, a scene that would be little changed since its abandonment by the bronze age people. The hillside itself has numerous remains from a comparable epoch. These include barrows, hut circles, holed stones, Neolithic fields and menhirs.

Tregeseal East circle looking north with Carn Kenidjack rising above it.

Once you reach the circles it can be seen that only one remains out of a possible three. The last remaining circle is known as Tregeseal East and the stones it contains have being knocked over and replaced many times. Most recently damage from grazing cattle using the stones as rubbing posts have caused a number to fall and be replaced. But even in 1893 the circles particularly the western one was described as being in a dilapidated state with the Mr Tregelles saying they were found as a "miserable wreck" (Cornishman 10th August 1893). The other circles laid on a West to East alignment with the eastern one, leading the eye towards the sea gap in the background.

Tregeseal through time

Looking through historic records, we can discover that the number of stones at the Tregeseal circles have been hard to decipher. William Borlase wrote in 1738: "On Tregeseal downs are two circles of stones placed on end, standing east and west of each other. In the eastern. 17 stones are still standing, two prostrate, one broken off. Diameter, 23 paces. In the western, 10 standing, four prostate, about 26 paces diameter, called Tregeseal dancing stones". With the earliest documentation recording Tregeseal East as having 19 stones, there is an unknown amount in the central circle and western circle. Although it should be noted that both these circles seen to be similar in size so could in theory have contained a similar amount of stones.
Both sites have been drawn through history by a number of artisans and archaeologists.

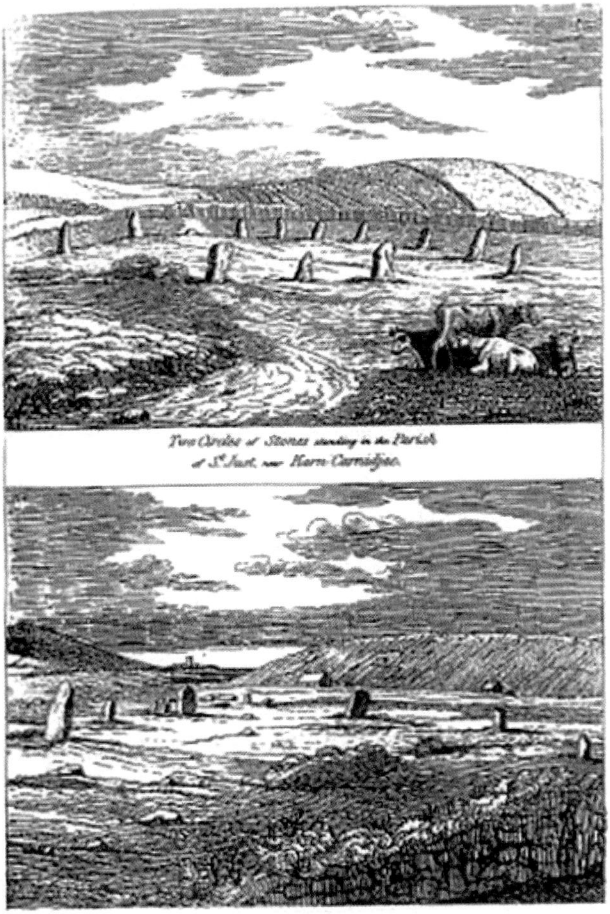

W.C. Cottons drawing from 1827. *Illustrations of stone circles, cromlechs, and other remains of the aboriginal Britons, in the West of Cornwall: from drawings made on the spot, in 1826.*

The top sketch shows Tregeseal East in quite a good state of repair, with at least 14 stones and gaps for the missing. The bottom image shows Tregeseal West circle showing St Just Church tower, The sea gap and The Isles of Scilly in the background. The number of stones are difficult to decipher from this image due to the wall and the amount of stones which seem to be dotted about in the field, but a figure of 8 stones seem to in the drawing. So, just shy of 90 years after Borlase had visited the site Cottons sketches show a deterioration in the stones in both circles.

Cotton also made the first plan of the site. It shows how the two circles are posited on an almost west to east cardinal alignment. The circle to the west has a attached cist similar to that at Boskednan and Boscawen-ûn.

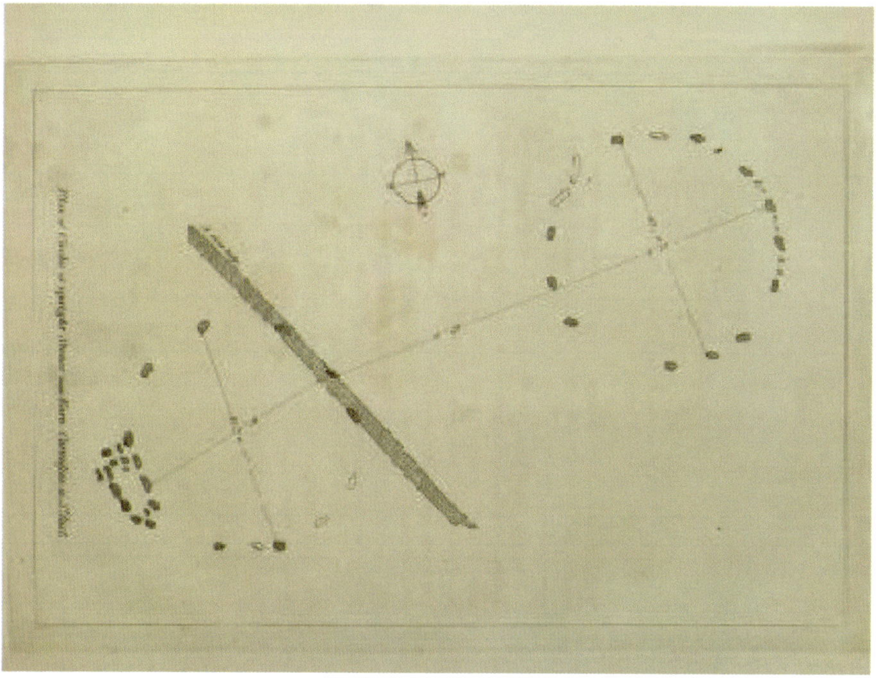

William Cotton 1827

Halgolluir

One of the first astronomical comments was made about the circles by John Butler in 1842 within
'Statistical Account of the Parish of St. Just: In Penwith, in the County of Cornwall'
He calls the circle complex Halgolluir after the nearby farmstead. He translates this to the 'down of the holy Moon. Hal for 'down', gol for 'holy' and luir for 'Moon'. He goes onto say "And here are two circles of upright stones: one, A dedicated to the Sun and B, which stands in the croft, and gives its name to the farm, dedicated to the Moon.' It is a little unclear why he dedicates one circle to the Sun. But the names link to the Moon is fascinating.
Interestingly he goes onto comment about the meaning of the word Carnyorth, saying that this translates to Carn of the Bear. Arth or Orth meaning Bear. With the western side of the hill being Carnbean Carnyorth or Carn of the Little Bear. He reflects that these could be directly linked to the constellations of the Ursa Major and Ursa Minor.
From his description, he is naming the region to the east of Carn Kenidjack as the 'Carn of the Bear' and Carn Bean as 'Carn of the little Bear'. Both are within a northern direction to the circles, but we should consider that the constellations were possibly not named or grouped as we see them currently. Ursa Major was first recognised as a constellation in Ptolemy's Almagest in the 2nd century BC, although Bradley E. Schaefer has suggested that the constellation was grouped as a bear for a much longer period of prehistory stretching back to 13000BC. Although the Chinese saw the part of the bear which makes up the plough as a chariot. But John Butlers translations are interesting nevertheless.
The final reference he gives in the region is another lunar one. He mentions that the hill further to the west of Carn Bean was known as "Top and dry Carn" or "Tuban druy Cann" which translates to "the bank of the druids full 'Moon'" . He then mentions that a circle of

stones from this hill had been carried off recently to Penzance. This mention of a further stone circle should be taken with caution as many of his prior circle of stones relate to barrow and hut remains more so than stone circles.
Butler draws the Tregeseal complex or Halgolluir with the two circles, showing the central circle, which is in partially incorporated in the field known as site B and the eastern circle known as site A.

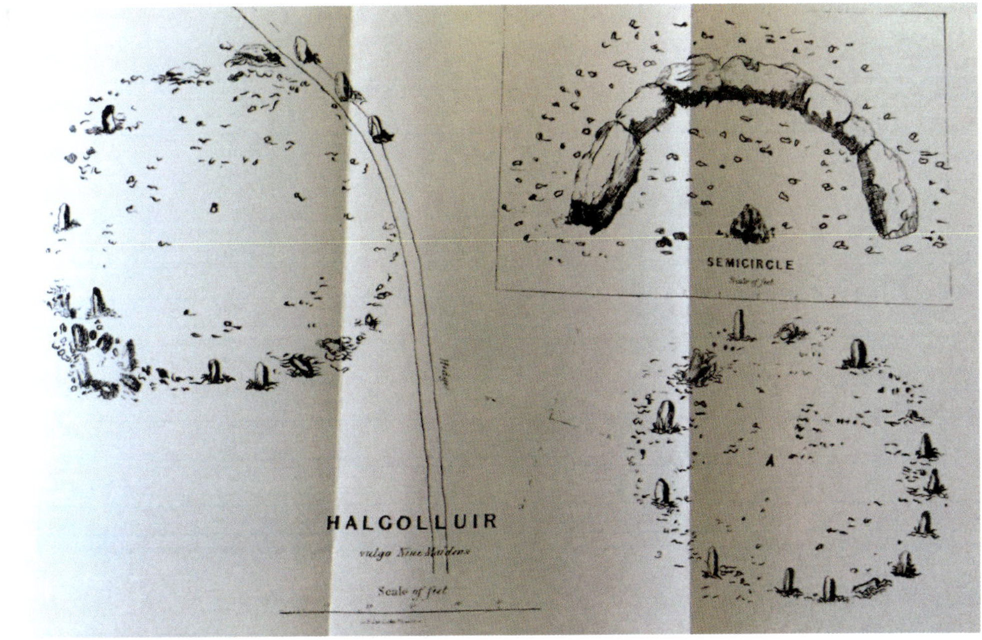

John Butler (1842) Statistical Account of the Parish of St. Just: In Penwith, in the County of Cornwall.

Richard Edmunds in 1862 considered that all the remaining circles (including Tregeseal central) in West Penwith had 19 stones, this he said, is a reflection of the metonic cycle. But the documentation about the Tregeseal circles through history is at best patchy and at worst is incorrect. It would be difficult to set upon an actual figure for the number of stones at the site. It is worth consideration that archaeological exploration of the area containing the best remaining example circle may also struggle to get to a conclusion as the site has been radically changed through the ages through quarrying and mining. Therefore, the actual number of stones contained in these circles is likely to remain a mystery.
William Copeland Borlase sketched a very simple plan of Tregeseal East in 1872. It shows that there are 16 stones and one which was labelled as missing.

William Copeland Borlase 1872

Borlase also took people out to visit the site and one of these guided walks was written about in the local paper, "The Cornish Telegraph". This walk W.C. Borlase led in early October 1880. It mentions the holed stones which are found to the NE of the circles and could well be linked with the whole complex astronomically. These holed stones are 5 in total and will be discussed later in this section, but are unusual and their purpose can only be theorised and not decisively deduced.

"From Tregeseal Mr. Borlase led the way across toe common to two stone circles, in the neighbourhood of which many chipped flints had been found; thou to a large barrow in which the late Mr. Butler found two stone axles and Mr. Borlase a third; and to two holed stones, which lay due west and were believed have been connected with the worship of the Sun. Mr. Borlase mentioned that in the highlands of Scotland there are similar holed stones through which lovers shake hands as pledge of betrothal."

Another early plan shows Tregeseal Central on the left and Tregeseal East on the right. This appears in the Victoria County History book and is interesting as it identifies a number of stones incorporated in the stone wall as part of Tregeseal Central. It also shows a much dilapidated Tregeseal East. Remember it is around this time that Mr Tregelles described the circles as being "miserable wrecks"

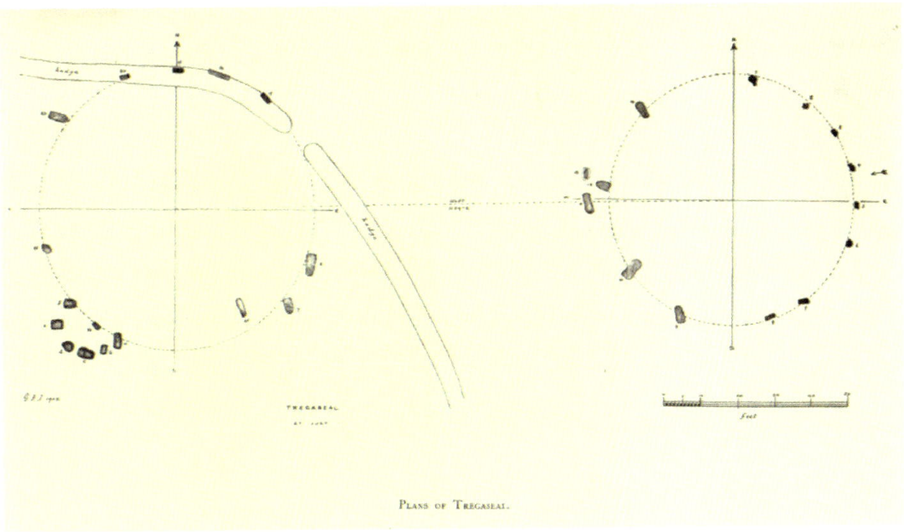

1902 Plan of Tregeseal stone circles from Victoria County History

In 1906 Norman Lockyer visited the site. Lady Lockyer photographed the eastern circle and it was published in the journal "Nature" and her husband's book "Stonehenge and other British Stone Monuments - Astronomically Considered". This image taken is as follows

Lady Lockyer photographs Tregeseal - Nature magazine 1906 Vol73

The image shows the landscape to the north-west, it shows Carn Kenidjack rising up in the background behind the circle.

What historic alignment suggestions did Lockyer suggest? He concentrated on the eastern circle and suggested six alignments. They were all the the NE of the site as the map below shows.

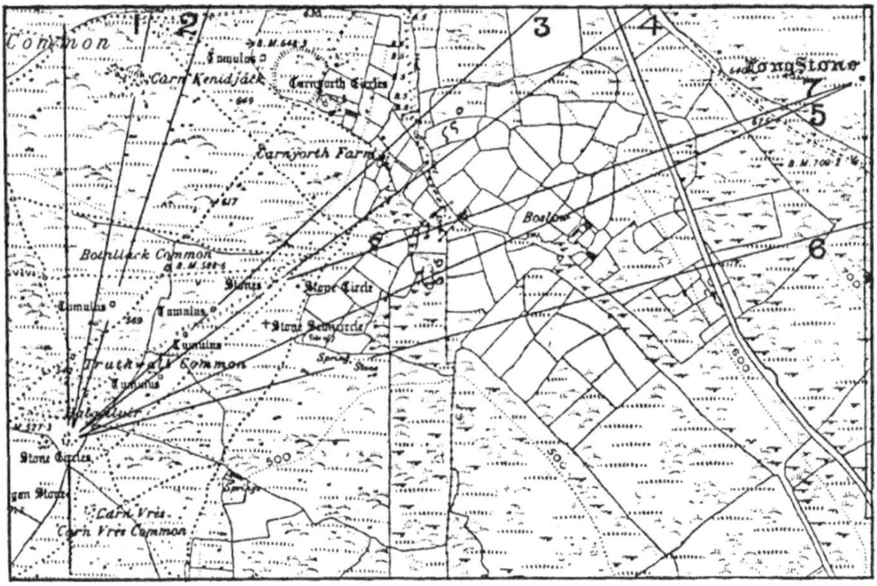

They are as follows

Alignments.	Azimuth.	Elevation.
1. Apex of Carn	N. 12° 8′ E.	4° 0′
2. Barrow 800′ distant	N. 20 8 E.	3 50
3. Two barrows 900′ distant	N. 50 3 E.	1 50
4. Holed stones	N. 53 20 E.	1 15
5. Longstone	N. 66 38 E.	2 10
6. Stone	N. 76 13 E.	

	Decl. N.	Star.	Date.
1. Apex of Carn	42° 33′ 0″	Arcturus	2330 B.C.
2. Barrow 800′ distant	40 29 0	,,	1970 ,,
3. Two barrows 900′ distant	25 20 21	? Solstitial	
4. Holed stones	23 2 20	? ,,	
5. Longstone	16 2 0	May sun	
6. Stone	9 15 0	Pleiades	1270 B.C.

All the targets were to the NE of the site,
1. Carn Kenidjack – the dominant and localised hill.
2.and 3 are local Barrows
4. Holed stones which may have moved
5. Longstone (Boswens menhir) - at distance but of importance as can see a large number of sites from here.
6. local stone

It is worth saying a little about the alignments suggested by Lockyer. Three are stellar alignments that are hard to prove without accurate dating of a site. One is to the cross-quarter mayday sun and two are to the summer solstice. Of these, the summer solstice alignments are perhaps the most credible suggestions. He does not consider lunar alignments within his study of the circle.

These alignments were given local newspaper coverage within the Royal Cornwall Gazette on the 8th May.

"Dealing with the circles at Tregeseal, not far from St. Just, Sir Norman says' and his wife visited the region in January this year, but previously to their going Mr. Horton Bolitho, accompanied by Mr. Thomas, whose knowledge of the local antiquities very great, had explored the region and taught them what to observe. The chief interest to lie the N.E. quadrant, where, in addition famous longstone on a hill about a mile away, the nest of holed stones and several the burrows were located. Cam Kenidjack. a famous landmark, lies to the north, of the two circles, sir Norman confined his attention almost exclusively to the eastern one, as the other was in a fragmentary condition, though it was still traceable. Mr. Horton Bolitho, who accompanied them 1st in January, had again visited the spot, with Mr. Thomas, for the purpose of further exploration, and determining the angular height of the skyline along the different alignments, which had plotted from the 6in. to the 25in. map. Sir Norman gives the approximate azimuths determined by himself therefrom. Of the sight-lines studied so far, those to and from the Longstone and the holed stones seem the most important. The Longstone, 1 1/2 miles the N.E., is a monolith 10ft. high on the western side of a hill; it is visible from the circle, though; furze has grown round and partly hidden it. He gives the meanings of the various alignments as they seem to him, and goes on to say that regarding the possible solstial alignments, the declinations obtained may neglected until the azimuth and angular heights of the hill have been determined with a good theodolite. A change —10' in the angular declination, and hence about that in the resulting declination, would bring the date given by the burrows to about 2.000 B.C...."

After Lockyer's initial work at Tregeseal and the other local circles at Lockyer's bequest local interested people formed "The Society for the Astronomical Study of Stone Monuments". This was the first organised archaeoastronomy group in the world and would for a number of years prove popular. Annual visits would be made to the stone circles in turn and in 1908 The Society for the Astronomical Study of Stone Monuments made their annual visit to Tregeseal. This was a culmination of further work which had been undertaken on site by group members with a theodolite

A further comprehensive report was written by *The Society for the Astronomical Study of Stone Monuments*

A report of their visit appeared in the Cornishman on the 2nd July 1908
"On the 21st December, 1906, the morning was very clear and the Sun rose just deg. sec. west of the barrow for that occasion in Botrea Downs. On the following anniversary of that date the weather was gloomy, so that the Sun could not be seen; but careful watching took place to ascertain the day which the Sun would actually rise on the barrow, and on the 6th of January, 1908, the Sun appeared directly on the barrow. On reference to Whittaker's almanack was found that the Sun's declination south at noon was reduced during that period by 49 sees.; but unfortunately all attempts witnessing a corresponding Sunset on

either of those days had failed. On the 4th of⍰February, 1908, the Sunrise was very clear, and⍰appeared to rise just 1 deg. 15 sec, south of the⍰barrow in Botrea Downs for that occasion. The⍰corresponding Sunset for that day was watched for,⍰but owing to cloud in the horizon, no minute⍰observation could be made. It, however, set near⍰the barrow in Tregeseal Croft. There is no existing⍰barrow that might be assigned for the rising⍰Sun of March 21, but not far from the little Dry⍰Cam, near the road there is evidence of something⍰ of the kind having once existed, and on that date⍰1908, the rising Sun was witnessed and appeared⍰somewhere about that spot. On the 6th May,⍰1908, the sky was cloudy in the morning, but in⍰the evening the Sunset was very bright to the last,⍰and it took place just 30 deg. 40 sec. north of the⍰barrow Truthwall Common. The following⍰morning very fine Sunrise was witnessed⍰when the Sun rose 1 deg.⍰15 sec. north of the Long Stone, and on the 19th⍰June, the Sun rose 3 deg. 40 sec. south of the two⍰barrows. The foregoing observations may be relied⍰on fairly correct, but when opportunity arises⍰they will repeated to test the readings of the⍰instrument. The evening observations so far have⍰all been of cursory character except that of the⍰Sunset of May 6, but enough has been seen to lead⍰to the conviction that objects for the various days⍰are 21st December, somewhere near the church,⍰ perhaps the old cross; Feb. 4, Tregeseal Croft;⍰March 21, Goon Reath; May 6, Truthwall Common;⍰whilst that of June 21 seems to be demolished⍰bringing the land into cultivation.⍰ "

The suggested alignments are as follows -all of them suggest local features with the solar cycle.

21st Dec 1906 Sunrise 1-degree west of Barrow on Botrese Downs – which is the winter solstice position

6th Jan 1908 Sunrise over the barrow

4th Feb 1908 Sunrise 1 degree 15 sec south of barrow

4th Feb 1908 Sunset near a barrow on Tregeseal Croft.

21st March Sunrise over dry cam near the road- the equinox position.

6th May 1908 Sunset 30-degree 40 sec north of barrow on Truthwall Common

7th May 1908 Sunrise 1 degree 15 sec north of the Longstone

19th June 1908 Sunrise 3-degree 40 sec south of two barrows

Of the groups suggestions the winter solstice position is interesting as is the equinox position. The others seem to have given current measurements to local monuments.

Aubrey Burl mentions that the circles lie east to west saying that the east-west positioning of the circles may have a crude relationship with the equinoxes. But even this seems unlikely as the circles are on a more NNE-SSW orientation.

Current Measurements

Recent onsite measurements have been taken and these have a list of local sites and natural site lines as follows

site	Azimuth	distance	altitude	visible
Higher Boslow stone circle?	38.09	695.31	3	N
Boslow barrow	57.48	569.65	1.7	N
Botrea Cairn 1	117.49	1840	1.1	N
Balleswidden barrow	149.66	1410	0.9	N
Carn Kenidjack (west)	7.44	559.29	3.8	Y
Carn Kenidjack (east)	14.08	604.38	3.8	Y
Truthwall common barrow 1	17.48	231.13	3.5	Y
Carnyorth common barrow	27.63	719.09	3.1	Y
Truthwall common barrow 2	47.63	268.37	2.2	Y
Truthwall common barrow 3	47.63	320.91	2.2	Y
Botollack common Holed stones	53.66	419.2	2	Y
Truthwall common barrow 4	61.41	223.69	1.4	Y
Truthwall common Cairn	61.41	385.29	1.4	Y
Boswens Standing stone	66.35	1440	1.8	Y
Boslow barrow 2	73.28	889.29	2.1	Y
Boslow Barrow 4	76.98	1180	2.1	Y
Dry Cairn barrow	79.47	1442	2.1	Y
Botrea Cairn 2	121.32	1845	1.4	Y
Botrea downs barrow	121.64	1841	1.4	Y
Botrea Cairn 3	125.32	1849	1.5	Y
Botrea Cairn 4	125.33	1853	1.5	Y
Botrea Cairn 5	127.68	1842	1.5	Y
Carn Vres barrow 1	139.44	101.39	1	Y
Carn Vres barrow 2	182.47	140.43	0.5	Y
Bostraze barrow	205.68	309.03	0	Y
Sea gap south edge	225	3520	0	Y
Isle of Scilly South	230	47000	0	Y
Isles of Scilly North	242	47000	0	Y
Tregeseal entrance grave	246.83	656.57	-0.3	Y
Sea gap north edge	248	3520	0	Y
Kenython barrow	267.01	545.09	0.6	Y
Truthwall common barrow 5	291.8	512.88	0.7	Y
Carn Beam Cairn 1 and 2	334.14	874.39	2.8	Y

All these sites are either visible from the site or they are within 2km from the site. The sea gap and the enclosed Isles of Scilly are a range with the north and south edges given.
Of these a number are visible on the horizon

site	Azimuth	distance	altitude	Declination	Horizon
Carn Kenidjack (west)	7.44	559.29	3.8	43.24	H
Carn Kenidjack (east)	14.08	604.38	3.8	42.25	H
Truthwall common barrow 1	17.48	231.13	3.5	41.08	H
Truthwall common barrow 2	48.63	268.37	2.2	26.92	H
Truthwall common barrow 3	48.88	320.91	2.2	26.79	H
Botollack common Holed stones	53.66	419.2	2	23.98	H
Boswens Standing stone	66.35	1440	1.8	15.58	H
Carn Vres barrow 1	140.98	101.39	1	-28.99	H
Carn Vres barrow 2	182.47	140.43	0.5	-39.32	H
Sea gap south edge	225	3520	0	-26.95	H
Isle of Scilly south edge	230	47000	0	-24.33	H
Isles of Scilly north edge	242	47000	0	-17.51	H
Sea gap north edge	248	3520	0	-13.89	H
Truthwall common barrow 5	291.8	512.88	0.7	14.34	H
Carn Beam Cairn 1 and 2	334.14	874.39	2.8	37.85	H

The sea gap has been measured as running between 225° to 248°. This is a large sea gap which frames the Isles of Scilly. They run through azimuth of 242° to 248°. The Isles of Scilly are visible at a smaller range within this sea gap. It is thought provoking to note that the winter solstice sunset falls within this sea gap range, as does the lunar minor southern setting position.

There are a number of sites which highlight the northern landscape alongside the Carn Kenidjack. Barrow, Cairns and also a bronze age settlement is in this direction. Towards the NW is Carn Beam with a number of Cairns at the suit. Interestingly the southern point has also been marked by Carn Vres barrow 2. There does seem to be an eye for the cardinal points with the east-west relationship of the circles and the north emphasised and southern position marked.

Carn Vres barrow 1 is in the position of the lunar major southern setting position. The barrow is found at a declination of -28.99°, the lunar standstill position is -29.85°.

In the direction of the summer solstice sunrise are two of the Truthwall common barrows, these are number barrow 1 and barrow 2.

Barrow 1 is found at a declination of 26.92° and barrow 2 is found at a declination of 26.79°. This is higher than the declination of summer solstice Sunrise in 2500BC in from the eastern circle. This barrow was imaged from the eastern circle on the summer solstice in 2017. The rising Sun in 2017 would be about two solar widths further south (right in the image) from where it would have risen in 2500BC. The barrow is on the horizon in the centre of the image. It is not in an exact alignment with the eastern circle and summer solstice Sunrise. What is interesting though is if you move to the location of the centre circle the barrow falls at an azimuth of 53.25° which places the closest Truthwall barrow at a declination of 23.39° which is in line with the summer solstice Sunrise.

The Holed Stones

The holed stones are at a declination of 23.98° which is very close to the summer solstice rising position. Could these holes stones have been used to measure the rising Sun? As a tool it is difficult to know whether this would have been the case. The stones themselves may have been moved and the holes have possibly changed their orientation. This positioning on the summer solstice Sunrise line from the eastern circle does suggest a connection to the Sun. It is worth asking that Individually and not with the circle could they have been used as a type of timekeeper over the course of the year?
Theoretically the holed stones could have been used to spot different positions on the eastern horizon and the rising Sun. Practically this could have worked in a similar manner to a pinhole camera. The observer would not need to lay uncomfortably on the floor, risking blindness looking directly at the Sun. Instead they could have used the shadow of the stones and waited for a perfect shadow hole to form to understand when the stone was in line with the rising eastern Sun. By lining up the stones with certain key dates they could safely be able to observe when that date was reached using this method. Alternatively, this could have also been used to spot different stars or the Moon within its cycle. Although we can never be sure that this is what these holed stones was used for a holed stone within Cornwall located at the Trethevy Quoit certainly had a link to the solar cycle. This holed stone is high off the ground located on the top of the capstone. Only when the Sun rises high enough around midday on the summer solstice period does it reach an altitude where it can shine directly through the hole to the ground below.

The holed stones on Botollack common

Boswens Longstone

Boswens standing stone was placed in its position to line up with the May Day Sunrise according to Lockyer. There is no reason to suspect this to be the case. At a declination of 15.58° from the eastern circle it does not have a relationship to either the solstice or equinox position. If it is measured from the central circle it has a declination of 15.72°. It is therefore also slightly out of any lunar extreme.
It is worth considering the accuracy of the suggestion of a May Sunrise, the start of May would have had a declination of 15.78°. This means it is in a accurate position for the central circle but too high for the eastern circle. Meaning that the Sun would have already passed this marker.

Orion's belt and triple circles.

Proving that ancient sites have been laid out to reflect what we see in the sky is fraught with complexity and pitfalls. Mankind likes to see patterns and linking patterns in the sky to what we see on the ground are quite often considered viable. But we must reflect that stars over time move in the sky, so the exact position of a star which we see now may not be the position which it was found in history. Also, we cannot know which stars or patterns an ancient culture would revere and even if they would immortalise them in stone.
Constellations are manmade patterns devised to group stars together often accompanied by

myths and legends. The stars which we as a society group together have not always been placed together in that manner in history. For instance, ancient societies in China has a completely separate group of constellations to that of the ancient Greeks.
Orion's belt is part of a long standing ancient Greek constellation of Orion. Even now Orion is perhaps one of the first constellations people will learn in the night sky particularly as it is easy to pick out the three belt stars Alnitak, Alnilam and Mintaka. The stars which make this distinctive pattern in the sky have been linked to a number of ancient sites including the pyramids and the triple stone circle at the Hurlers in Bodmin Moor. The three stars are bright and part of a distinctive pattern rising high in the sky during the long winter nights. Mintaka currently rises at an almost eastern position 90° and sets at an almost western position 270° making it an important star for navigational purposes. But in 2500BC Mintaka did not hold that eastern rising position. It would have risen at 113°. A suggested alternative for a star being on this cardinal position in the bronze age is the nearby Pleiades cluster. It holds this carinal rising and setting position in 2500BC, rising in the east at 89° and setting at 272°.
Proving that the layout of the circle is linked to the stars in the sky would be very difficult without corroborative evidence, such as written records to back this up. The site at Tregeseal is in a dilapidated state and it is unknown whether the third circle ever exist acted. It would be especially difficult to say convincingly that the site was laid out to reflect a star pattern in the sky.

Processional routes for the winter solstice.

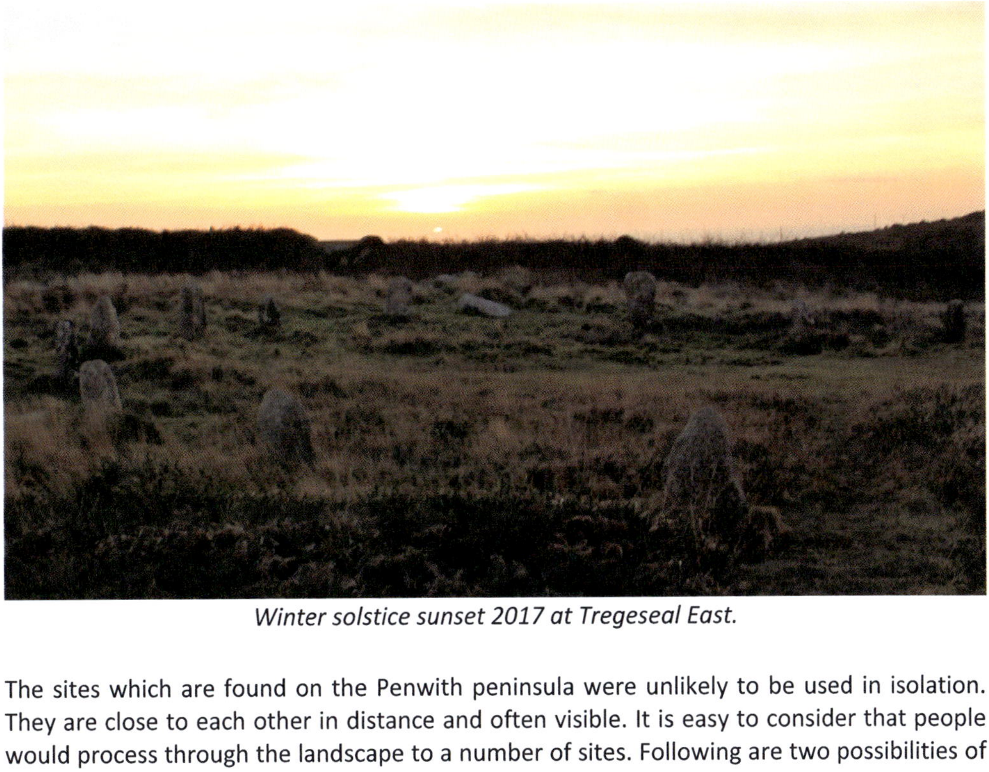

Winter solstice sunset 2017 at Tregeseal East.

The sites which are found on the Penwith peninsula were unlikely to be used in isolation. They are close to each other in distance and often visible. It is easy to consider that people would process through the landscape to a number of sites. Following are two possibilities of processional routes they both fall on the winter solstice the first includes the Boscawen-ûn circle site to Tregeseal and the second includes Boskednan to the Tregeseal complex.

The first procession could have occurred through the southern landscape on winter solstice. A group of people could have processed from the Lamorna gap using the Trelew line of stones to Boscawen-ûn where they could have witnessed the winter solstice Sunrise out of the gap. It would be possible for them to make their way through a ceremonial landscape perhaps visiting the framed Chapel Carn Brea next. From there the Longstone at Boswens is visible, they could have used this to parade to the Tregeseal complex where the winter solstice Sunset would have occurred in another sea gap over the Isles of Scilly.
Proving that these sites were used in this manner would be difficult, but by having the winter solstice Sunrise in a sea gap at one circle site and the winter solstice Sunset at another would suggest that they could have been linked in this way.

The second procession could have occurred through the northern landscape with people processing from Boskednan to Tregeseal. This would use the Boswen's menhir which falls close to the winter solstice setting position as a marker on the landscape where people would be able then to approach the Tregeseal complex to view the winter solstice sunset over the Isles of Scilly through the sea gap.

Visibility between the sites

When considering inter-visibility there are some considerations, one is the practical side and how far can people realistically see? and the other is why they would have wanted to see over large distances.
Practically we have to consider what detail can we see as well over long distances, as the curvature of the Earth. The curvature of the Earth would not be an issue on the Penwith peninsula, but distance and eyesight would certainly be a limiting factor. The use of fire at night could have lit up sites making them more visible to each other over long distances, but in the daytime how far we can see is limited not only by our eyesight but the weather conditions on the day.
The reason as to why they want sites to be indivisibly could be multiple. These could include but are not exhaustive reasons of ceremonial, practical or even economic reason.
Tregeseal is the only remaining circle site which is not visible from any of the other circle sites. But the Longstone at Boswens is perhaps a link that makes it inter-visible to Boskednan.
Boscawen-ûn is visible to the Merry Maidens complex and over very long distances the Merry Maidens can be seen from Boskednan. It would be impossible to say whether this was part of a grand design. There are many other circle sites which have now disappeared from the landscape. It seems that the inter-visibility of the sites as a whole seems to be less of a factor than the positioning in the local landscape to highlight nearby features such as Cairns. The exception to this seems to be the Merry Maidens complex. The Merry Maidens complex is directly south from the distant Boskednan circle and it is also in the lunar major southern rising position from Boscawen-ûn.

Conclusions

The following table has several suggested alignments which have been discussed for each of the circles. There are four highlighted in blue and these are what this study considers the likely main focus alignment of the circle. Reading the table, the azimuth locations start in the north and work clockwise around, moving through the rising positions before reaching the setting positions and back towards the north.

	Boscawen-ún	Boskednan	Merry M.	Tregeseal East	Total
North (or close)	Local Hill	Bosporthennis	Carn Galver	Local ridge	4
Lunar Maj Northern	Field Menhir	Zennor Ridge	Pipers		3
Summer Solstice	Hedge Menhir			Truthwall Bar.	2
Lunar Min Northern					0
East			Rosm. holed	Barrow	2
Lunar Min Southern					0
Winter Solstice	Lamorna Gap	Sea Gap			2
Lunar Maj Southern	Merry Maidens				1
South		Merry Maidens	Local Ridge		2
Lunar Maj Southern				Carn Vres Bar.	1
Winter Solstice		Boswens Menhir		Isles of Scilly	2
Lunar Min Southern				Sea Gap	1
West					0
Lunar Min Northern					0
Summer Solstice	Barrow				1
Lunar Maj Northern	Creeg Tol	Carn Galver	Bartinney		3
Total	7	6	5	6	24

Further analysis can split these out into types of alignments

Type	Found	Possible
Cardinal	8	16
Lunar Maj Northern	6	8
Lunar Maj Southern	2	8
Lunar Min Northern	0	8
Lunar Min Southern	1	8
Summer Solstice	3	8
Winter Solstice	4	8
Total	24	64

Of the circles that remain three of the four have a northern localised hill. This may not be the most dominant landscape feature, but these also tend to be positioned just of the northern cardinal point. For example, the midpoint of the Carn Kenidjack rocky outcrop is just off north at the Tregeseal circles at 11° but the ridge between Carn Kenidjack and Carn

Bream which runs through the northern most azimuth is still one of the highest positions on the horizon.
Of this theme of northern hills near circles the Merry Maidens (and its neighboring circle Boleigh) are the exceptions, as they have a southern ridge nearby and large expansive views to the north. They are also unusual as having no sea gaps visible from the site.
What is interesting is that circles having this dominant feature to the north have chosen a part of the horizon to feature where the Sun cannot rise or set. Also, it may be worth consideration that the dominant features of Creeg Tol at Boscawen-ún and Carn Galver at Boskednan are also at the limit of where the Moon can set in a once in 19-year event.
Did the circles get positioned so the local dominant hill cannot be touched by the Sun, (or very unlikely the Moon)? This could be the case as it would also be true for circles with a dominating southern feature. They may have wanted to choose an area to highlight which was unaffected by the rising or setting Sun. The northern hills could equally be positioned to highlight the circumpolar stars surrounding the north celestial pole. Although, other non-astronomical reasons may also explain this relationship with circles and an elevated northern horizon.
As explained within the book some of these dominant northern features are not positioned directly at the northern cardinal position. For example, Carn Kenidjack is for just off the northern Cardinal point. Further afield similar azimuths are found at circles with dominant hills towards to the north around the country. Which could lead to a conclusion that they were aligning something more particular such as a star, or the limits of th lunar cycle.
Let's add I some evidence from farther afield where northern features are regularly found Cardinal points are found on Bodmin Moor where, Fernacre circle is due south from Rough Tor and north of Garrow Tor. At the Stripple stones you can see Garrow Tor to the north and then the Rough Tor beyond that. Also from The Hurlers stone circles the Cheesewring is 4° west of north with Stowe Pound in a general northern direction they also have Rillaton barrow at 14° east of north as measured from the central circle. Similar to Tregeseal and it position to Carn Kenidjack you can find a line from the Tripplet stone circle through Leaze circle to Rough Tor at 11° east of north.
We could consider that the grouping of alignments close to 11° of north, they could relate to a rising a setting star such as Arcturus. This star would have risen with a declination of 42°47'42''and a azimuth of about 6° at a 0 altitude in 2000BC.
There is also a grouping of alignments around the lunar northern standstill position at 33° and 34°. Once more several examples of these can be found from further afield in Bodmin Moor. Like Carn Galver's position from Boskednan, we find alignments with the lunar major northern standstill positions at the Stripple stones circle where, Hawkstor is the dominant hill is 34° west of north and there is another hill 34° east of north. At Leaze circle Garrow Tor is found 33° east of north. This azimuth has also been found when looking at the alignments between circles. At Stanton Drew in Somerset the northern circle is 35° east of north from the southern circle. And even farther afield a circle at Keswick in Cumbria has Skiddaw and Blencathra hills 35° east and west of north respectively.
It seems that there is a lot of evidence for marking the major northern lunar standstill position, with the four circles examined in the book, offering six possible alignments out of eight. Why would they have chosen the northern standstill position and not the other standstill positions? They would possibly have had a better opportunity to mark this position. The full Moon reaches its most northerly azimuth rising and setting position nearest the

winter solstice. It also would have been rising and setting to its highest altitude making a real show in the dark nights of winter.

Marking the position of this mid-winter full moon on the horizon would eventually lead to an almost perfect alignment of the northern major standstill position. Although this would also be possible to do in the summer with the full moon nearest the summer solstice we have to remember that the moon would not be as impressive in appearance as it would have still been rising and setting in the daylight and it would rise to its lowest altitudes during the year. It seems that there is a real case for the alignment of the northern major standstill positions at Boskednan and Boscawen-ún. The northern standstill alignments at the Merry Maidens are different in nature (distant hill, possible menhir alignment) but there is no reason to think that this is not a real attempt for the people in West Penwith trying to ritualise the lunar standstill positions within the circles. The number of lunar alignments was perhaps the most surprising discovery of this project. It had been expected that several the circles would have solstice alignments as these are well documented around the UK but the knowledge and positioning of the circles for the motion of the moon is, although not unique, was perhaps surprising. Although we must remember that these people would have had a strong relationship and reliance with the sea for trading, transport and food. Therefore, it is no stretch to believe that they would have had a detailed knowledge of the motions of the tides and therefore a real interest in the motion of the Moon.

Full Moon closest to winter solstice from Mên-an-Tol rising above Boskednan (2017)

It is worth noting that solstice positions have been marked on more than one occasion with a rising or setting sun from a sea gap. Anyone who has watched a sunrise or sunset over the sea can appreciate how special this event can be. To frame the extremes of the cycle in with a solstice sun with the ocean continues to highlight the relationship that these people had with the sea surrounding them.

Of course, within West Penwith this is not the whole story. There are many other circle sites which are known to have existed and even more which are now only documented within the archives and ancient texts. These sites include the following – (this is not an exhaustive list!).

<u>Near Boscawen-ún</u>

Higher Trevorian to the NW of St Buryan this site would have been visible to the south from Boscawen-ún.

<u>Near Merry Maidens</u>

Rosmodress circle (Tregurnow) to the east of the merry Maidens site near Boleigh

Boleigh circle in the field below the Merry Maidens

<u>Near Tregeseal</u>

Tregeseal the Western circle only found on aerial photographs as a crop mark it continues the line of the eastern and central circle to the west.

Botallack multi stone site – first mentioned by Borlase now no trace but in the region of Botallack mines

Circles above Tregeseal – Soldiers Croft

<u>Near Boskednan</u>

The Zennor Cirque site. – North of Mulfa Quoit on the eastern side of a localised ridge with a Southern Hill. This may have been a circle or an enclosure

Mên-an-Tol - Built east to the ridge with Boskednan circle. This has only few stones standing including the infamous holed stone.

Rosemergy – a possible circle site north of Watch Croft near the tea rooms

On a final note the whole peninsula is densely covered with ancient remains other sites many of these may also have relationships to astronomical events or celestial objects.

Index